Factors of Soil Formation:
A Fiftieth Anniversary Retrospective

Factors of Soil Formation: A Fiftieth Anniversary Retrospective

Proceedings of a symposium cosponsored by the Council on the History of Soil Science (S205.1) and Division S-5 of the Soil Science Society of America. The symposium was held in Denver, CO, 28 October 1991.

Organizing Committee
Ronald Amundson
John Tandarich

Editorial Committee
Ronald Amundson
Jennifer Harden
Michael Singer

Editor-in-Chief SSSA
Robert J. Luxmoore

Managing Editor
Jon M. Bartels

SSSA Special Publication Number 33

**Soil Science Society of America, Inc.
Madison, Wisconsin, USA**

1994

Cover Design: adapted from a design by Hans Jenny

Copyright © 1994 by the Soil Science Society of America, Inc.

ALL RIGHTS RESERVED UNDER THE U.S. COPYRIGHT
ACT OF 1976 (P.L. 94-553)

Any and all uses beyond the limitations of the "fair use" provision of the law require written permission from the publisher(s) and/or the author(s); not applicable to contributions prepared by officers or employees of the U.S. Government as part of their official duties.

Soil Science Society of America, Inc.
677 South Segoe Road, Madison, WI 53711 USA

Reprinted in 1994.

Library of Congress Cataloging-in-Publication Data

Factors of soil formation : a fiftieth anniversary retrospective proceedings of a symposium / sponsored by Division S-5 of the Soil Science Society of America ; organizing committee, Ronald Amundson, John Tandarich : editorial committee, Ronald Amundson, Jennifer Harden, Michael Singer.
     p.   cm. —— (SSSA special publication; no. 33)
"The symposium was held in Denver, CO, 28 October 1991."
Includes bibliographical references.
ISBN 0-89118-804-5
    1. Soil formation—Congresses. 2. Soils—Congresses. I. Amundson, Ronald (Ronald G.) II. Harden, J.W. (Jennifer Willa), 1954- . III. Singer, Michael J. (Michael John), 1945- . IV. Soil Science Society of America. Division S-5. V. Series.
S592.2.F33   1993
551.3'05—dc20   93-41400
                                                                       CIP

Printed in the United States of America

CONTENTS

	Page
Foreword	vii
Preface	ix
Contributors	xi
Conversion Factors for SI and non-SI Units	xiii

1 The Intellectual Background for the Factors of Soil Formation
    John P. Tandarich and Steven W. Sprecher 1

2 Factors of Soil Formation: Contributions to Pedology
    L. P. Wilding 15

3 The Environmental Factor Approach to the Interpretation of Paleosols
    Gregory J. Retallack 31

4 The "State Factor" Approach in Geoarchaeology
    Vance T. Holliday 65

5 Factors Controlling Ecosystem Structure and Function
    Peter M. Vitousek 87

6 Soil Geography and Factor Functionality: Interacting Concepts
    R. W. Arnold 99

7 Soil Formation Theory: A Summary of its Principal Impacts on Geography, Geomorphology, Soil-Geomorphology, Quaternary Geology, and Paleopedology
    D. L. Johnson and F. D. Hole 111

8 Towards a New Framework for Modeling the Soil-Landscape Continuum
    K. McSweeney, P. E. Gessler, B. Slater, R. D. Hammer, J. Bell, and G. W. Petersen 127

Appendix 1
    Memories of Professor Hans Jenny
        Rodney J. Arkley 147

Appendix 2
    We Remember Hans Jenny
        Gordon L. Huntington 149

Appendix 3
    Brief Highlights of Hans Jenny's Life, Publications of Hans Jenny
        Ronald Amundson 153

FOREWORD

The book *Factors of Soil Formation. A System of Quantitative Pedology* published in 1941 represented a quantum leap forward in our understanding of the impacts of parent material, climate, topography, vegetative cover and time on soil character. Even after more than 50 yr, Hans Jenny's elegant theory of soil formation remains the foundation of modern thought regarding soil genesis. This SSSA Special Publication includes papers presented at a special symposium commemorating the 50th anniversary of the publication of *Factors of Soil Formation* held during the 1991 annual meeting of the Soil Science Society of America. Also included are special tributes to Professor Jenny. As in many other things, we have much to learn from a retrospective look at the original contributions that make up the foundations of modern soil science.

DARRELL W. NELSON, *president*
Soil Science Society of America

Hans Jenny

PREFACE

In 1941, McGraw-Hill Book Company published *Factors of Soil Formation. A System of Quantitative Pedology*. One may ask, of what relevance is a 50 yr-old textbook to modern scientists? There are two responses to such a question. First, *Factors of Soil Formation. A System of Quantitative Pedology* is no textbook in the ordinary sense. It is from cover to cover a concise and detailed elaboration of a theory—how soils and (although Jenny did not use the word) ecosystems evolve. Second, the mere passage of time does not make a theory obsolete or dated; only a new theory can do that. As the scientific historian Thomas Kuhn (1970) has suggested, the replacement of one theory by another is not a trivial event, but it occurs so rarely, and has such significant scientific consequences, that it is said to constitute a scientific revolution.

As practicing scientists, we value the importance of theories, yet we seldom recognize how our collection of accepted theories and ideas dictate the way in which we perceive the world and conduct our research. Gunnar Myrdal (Simonson, 1991, p. 8) wrote, "Theory is necessary not only to organize the findings of research so that they make sense, but, more basically, to determine the questions to be asked. Theory, therefore, must always be *a priori* to the empirical observation of facts. Facts come to mean something only as ascertained and organized in the frame of a theory."

Theories and ideas are, therefore, the tools we use to understand the natural world. In the climactic scene of *Inherit the Wind*, Spencer Tracy, as the attorney Clarence Darrow arguing in the defense of the biology teacher John Scopes, exclaimed to the packed courtroom that "An idea is a greater monument than a cathedral." If that is true, then the best measure of any ideological monument, such as *Factors of Soil Formation. A System of Quantitative Pedology* is to determine how far, and how wide, its shadow is cast.

The papers in this special volume represent the views of a multidisciplinary group of authors as to how *Factors of Soil Formation. A System of Quantitative Pedology* has influenced scientific thought and research in their respective fields. The papers were first presented 28 Oct. 1991 at the annual meetings of the Soil Science Society of America, in the symposium commemorating the 50th anniversary of the publication of Professor Hans Jenny's seminal book. Special recognition of Professor Jenny is included as an addendum in this publication.

The symposium that led to this publication was cosponsored by the Soil Genesis, Morphology, and Classification Division (S-5) and the Council on the History of Soil Science of the Soil Science Society of America. We thank the authors for their participation in the symposium and for their contributions to this important addition to the history and philosophy of Pedology.

Organizing Committee

RONALD AMUNDSON
University of California
Berkeley, California

JOHN TANDARICH
Hey and Associates
Chicago, Illinois

Editorial Committee

RONALD AMUNDSON
Unifersity of California
Berkeley, California

JENNIFER HARDEN
U.S. Geological Service
Menlo Park, California

MICHAEL SINGER
University of California
Davis, California

REFERENCES

Kuhn, T.S. 1970. The structure of scientific revolutions. 2nd ed. Univ. Chicago Press, Chicago, IL.

Simonson, R.W. 1991. Soil science—goals for the next 75 years. SOil Sci. 151:7-18.

CONTRIBUTORS

Ronald Amundson — Associate Professor, Department of Soil Science, 108 Hilgard, University of California, Berkeley, California 94720

Rodney J. Arkley — Lecturer Emeritus, Department of Soil Science, 108 Hilgard, University of California, Berkeley, California 94720

R. W. Arnold — Director of the Soil Survey, USDA-SCS, P.O. Box 2890, Washington, District of Columbia 20013

J. C. Bell — Assistant Professor of Soil Science, Department of Soil Science, University of Minnesota, St. Paul, Minnesota 55108

P. E. Gessler — Australian National University, G.P.O. Box 639, Canberra, Australia

R. David Hammer — Associate Professor of Soil Science, School of Natural Resources, University of Missouri, Columbia, Missouri 65211

Francis D. Hole — Professor Emeritus of Soil Science and Geography, University of Wisconsin, Madison, Wisconsin 53706

Vance T. Holliday — Associate Professor of Geography, Department of Geography, University of Wisconsin, Madison, Wisconsin 53706

Gordon L. Huntington — Professor Emeritus of Soil Morphology and Genesis, Department of Land, Air, and Water Resources, University of California, Davis, California 95616

D. L. Johnson — Professor of Geography and Pedology, Department of Geography, University of Illinois, Urbana, Illinois 61801

K. McSweeney — Associate Professor of Soil Science, Department of Soil Science, University of Wisconsin, Madison, Wisconsin 53706

G. W. Petersen — Professor of Soil Genesis and Morphology, 116 A.S.I. Building, Pennsylvania State University, University Park, Pennsylvania 16802

Gregory J. Retallack — Professor of Geological Sciences, Department of Geological Sciences, University of Oregon, Eugene, Oregon 16802

B. K. Slater — Graduate Student, Department of Soil Science, University of Wisconsin, Madison, Wisconsin, 53706

Steven W. Sprecher — Pedologist, U.S. Army Corps of Engineers Waterways Experiment Station, Vicksburg, Mississippi 39180

John P. Tandarich — Soil Scientist and Archaeologist, Hey and Associates, Inc., 53 W. Jackson Blvd., Suite 1015, Chicago, Illinois 60604

Peter M. Vitousek Professor of Biological Sciences, Stanford University, Stanford, California 94305

L. P. Wilding Professor of Pedology, Soil and Crop Sciences Department, Texas A&M University, College Station, Texas 77843

Conversion Factors for SI and non-SI Units

Conversion Factors for SI and non-SI Units

To convert Column 1 into Column 2, multiply by	Column 1 SI Unit	Column 2 non-SI Unit	To convert Column 2 into Column 1, multiply by
Length			
0.621	kilometer, km (10^3 m)	mile, mi	1.609
1.094	meter, m	yard, yd	0.914
3.28	meter, m	foot, ft	0.304
1.0	micrometer, μm (10^{-6} m)	micron, μ	1.0
3.94×10^{-2}	millimeter, mm (10^{-3} m)	inch, in	25.4
10	nanometer, nm (10^{-9} m)	Angstrom, Å	0.1
Area			
2.47	hectare, ha	acre	0.405
247	square kilometer, km^2 (10^3 m)^2	acre	4.05×10^{-3}
0.386	square kilometer, km^2 (10^3 m)^2	square mile, mi^2	2.590
2.47×10^{-4}	square meter, m^2	acre	4.05×10^3
10.76	square meter, m^2	square foot, ft^2	9.29×10^{-2}
1.55×10^{-3}	square millimeter, mm^2 (10^{-3} m)^2	square inch, in^2	645
Volume			
9.73×10^{-3}	cubic meter, m^3	acre-inch	102.8
35.3	cubic meter, m^3	cubic foot, ft^3	2.83×10^{-2}
6.10×10^4	cubic meter, m^3	cubic inch, in^3	1.64×10^{-5}
2.84×10^{-2}	liter, L (10^{-3} m^3)	bushel, bu	35.24
1.057	liter, L (10^{-3} m^3)	quart (liquid), qt	0.946
3.53×10^{-2}	liter, L (10^{-3} m^3)	cubic foot, ft^3	28.3
0.265	liter, L (10^{-3} m^3)	gallon	3.78
33.78	liter, L (10^{-3} m^3)	ounce (fluid), oz	2.96×10^{-2}
2.11	liter, L (10^{-3} m^3)	pint (fluid), pt	0.473

CONVERSION FACTORS FOR SI AND NON-SI UNITS

Mass

To convert Column 1 into Column 2, multiply by	Column 1 SI Unit	Column 2 non-SI Unit	To convert Column 2 into Column 1, multiply by
2.20×10^{-3}	gram, g (10^{-3} kg)	pound, lb	454
3.52×10^{-2}	gram, g (10^{-3} kg)	ounce (avdp), oz	28.4
2.205	kilogram, kg	pound, lb	0.454
0.01	kilogram, kg	quintal (metric), q	100
1.10×10^{-3}	kilogram, kg	ton (2000 lb), ton	907
1.102	megagram, Mg (tonne)	ton (U.S.), ton	0.907
1.102	tonne, t	ton (U.S.), ton	0.907

Yield and Rate

0.893	kilogram per hectare, kg ha^{-1}	pound per acre, lb acre^{-1}	1.12
7.77×10^{-2}	kilogram per cubic meter, kg m^{-3}	pound per bushel, lb bu^{-1}	12.87
1.49×10^{-2}	kilogram per hectare, kg ha^{-1}	bushel per acre, 60 lb	67.19
1.59×10^{-2}	kilogram per hectare, kg ha^{-1}	bushel per acre, 56 lb	62.71
1.86×10^{-2}	kilogram per hectare, kg ha^{-1}	bushel per acre, 48 lb	53.75
0.107	liter per hectare, L ha^{-1}	gallon per acre	9.35
893	tonnes per hectare, t ha^{-1}	pound per acre, lb acre^{-1}	1.12×10^{-3}
893	megagram per hectare, Mg ha^{-1}	pound per acre, lb acre^{-1}	1.12×10^{-3}
0.446	megagram per hectare, Mg ha^{-1}	ton (2000 lb) per acre, ton acre^{-1}	2.24
2.24	meter per second, m s^{-1}	mile per hour	0.447

Specific Surface

10	square meter per kilogram, m^2 kg^{-1}	square centimeter per gram, cm^2 g^{-1}	0.1
1000	square meter per kilogram, m^2 kg^{-1}	square millimeter per gram, mm^2 g^{-1}	0.001

Pressure

9.90	megapascal, MPa (10^6 Pa)	atmosphere	0.101
10	megapascal, MPa (10^6 Pa)	bar	0.1
1.00	megagram per cubic meter, Mg m^{-3}	gram per cubic centimeter, g cm^{-3}	1.00
2.09×10^{-2}	pascal, Pa	pound per square foot, lb ft^{-2}	47.9
1.45×10^{-4}	pascal, Pa	pound per square inch, lb in^{-2}	6.90×10^3

(continued on next page)

Conversion Factors for SI and non-SI Units

To convert Column 1 into Column 2, multiply by	Column 1 SI Unit	Column 2 non-SI Unit	To convert Column 2 into Column 1, multiply by
		Temperature	
1.00 (K − 273)	Kelvin, K	Celsius, °C	1.00 (°C + 273)
(9/5 °C) + 32	Celsius, °C	Fahrenheit, °F	5/9 (°F − 32)
		Energy, Work, Quantity of Heat	
9.52×10^{-4}	joule, J	British thermal unit, Btu	1.05×10^{3}
0.239	joule, J	calorie, cal	4.19
10^{7}	joule, J	erg	10^{-7}
0.735	joule, J	foot-pound	1.36
2.387×10^{-5}	joule per square meter, J m^{-2}	calorie per square centimeter (langley)	4.19×10^{4}
10^{5}	newton, N	dyne	10^{-5}
1.43×10^{-3}	watt per square meter, W m^{-2}	calorie per square centimeter minute (irradiance), cal cm^{-2} min^{-1}	698
		Transpiration and Photosynthesis	
3.60×10^{-2}	milligram per square meter second, mg m^{-2} s^{-1}	gram per square decimeter hour, g dm^{-2} h^{-1}	27.8
5.56×10^{-3}	milligram (H$_2$O) per square meter second, mg m^{-2} s^{-1}	micromole (H$_2$O) per square centimeter second, μmol cm^{-2} s^{-1}	180
10^{-4}	milligram per square meter second, mg m^{-2} s^{-1}	milligram per square centimeter second, mg cm^{-2} s^{-1}	10^{4}
35.97	milligram per square meter second, mg m^{-2} s^{-1}	milligram per square decimeter hour, mg dm^{-2} h^{-1}	2.78×10^{-2}
		Plane Angle	
57.3	radian, rad	degrees (angle), °	1.75×10^{-2}

CONVERSION FACTORS FOR SI AND NON-SI UNITS

Electrical Conductivity, Electricity, and Magnetism

To convert Column 1 into Column 2, multiply by	Column 1 SI Unit	Column 2 non-SI Unit	To convert Column 2 into Column 1, multiply by
10	siemen per meter, S m^{-1}	millimho per centimeter, mmho cm^{-1}	0.1
10^4	tesla, T	gauss, G	10^{-4}

Water Measurement

9.73 × 10^{-3}	cubic meter, m^3	acre-inches, acre-in	102.8
9.81 × 10^{-3}	cubic meter per hour, m^3 h^{-1}	cubic feet per second, ft^3 s^{-1}	101.9
4.40	cubic meter per hour, m^3 h^{-1}	U.S. gallons per minute, gal min^{-1}	0.227
8.11	hectare-meters, ha-m	acre-feet, acre-ft	0.123
97.28	hectare-meters, ha-m	acre-inches, acre-in	1.03 × 10^{-2}
8.1 × 10^{-2}	hectare-centimeters, ha-cm	acre-feet, acre-ft	12.33

Concentrations

1	centimole per kilogram, cmol kg^{-1} (ion exchange capacity)	milliequivalents per 100 grams, meq 100 g^{-1}	1
0.1	gram per kilogram, g kg^{-1}	percent, %	10
1	milligram per kilogram, mg kg^{-1}	parts per million, ppm	1

Radioactivity

2.7 × 10^{-11}	becquerel, Bq	curie, Ci	3.7 × 10^{10}
2.7 × 10^{-2}	becquerel per kilogram, Bq kg^{-1}	picocurie per gram, pCi g^{-1}	37
100	gray, Gy (absorbed dose)	rad, rd	0.01
100	sievert, Sv (equivalent dose)	rem (roentgen equivalent man)	0.01

Plant Nutrient Conversion

	Elemental	Oxide	
2.29	P	P$_2$O$_5$	0.437
1.20	K	K$_2$O	0.830
1.39	Ca	CaO	0.715
1.66	Mg	MgO	0.602

1 The Intellectual Background for the Factors of Soil Formation

John P. Tandarich
Hey and Associates, Inc.
Chicago, Illinois

Stephen W. Sprecher
U.S. Army Corps of Engineers Waterways Experiment Station
Vicksburg, Mississippi

ABSTRACT

Modern pedology has developed from the emerging disciplines of seventeenth and eighteenth century chemistry, geography and geology, which in turn had roots in classical speculation about the nature of matter. It was during the nineteenth century that the idea slowly developed that soils are separate from other earth phenomena. They were thought by the geologists to form from the weathering of rocks and by the chemists to be the product of organic matter. The idea that several interrelated factors contribute to soil formation was possible only when soil was fully recognized as a unique natural body. This multifactor genesis of soils was most fully articulated in the nineteenth century by Vasilli Dokuchaev in Russia and Eugene Hilgard in the USA. While Hilgard developed his own ideas concerning the nature and formation of soils, it was not so clear how Dokuchaev thought of the idea of multiple factors of soil formation. Dokuchaev's inspiration may have come, at least in part, from the Russian chemist Dmitri Mendeleev. In 1876, Mendeleev suggested that the Imperial Free Economic Society of Saint Petersburg form a commission to study the black earth or chernoziom, and this commission consisted of representatives from agronomy, chemistry, geography, geology, physics and zoology. The commission chairman was Dokuchaev, and his thinking on the factors of soil formation must have crystallized using the interdisciplinary framework of the commission structure. While not quantified, the factors provided a basis for understanding soils which was carried to the West through Dokuchaev's student Konstantin Glinka in collaboration with the German pedologist Hermann Stremme to Curtis Marbut and Hans Jenny. Additionally, Jenny rediscovered and reintroduced Hilgard's independently conceived ideas of soil formation to the pedological community.

Copyright © 1994 Soil Science Society of America, 677 S. Segoe Rd., Madison, WI 53711, USA. *Factors of Soil Formation: A Fiftieth Anniversary Retrospective.* SSSA Special Publication 33.

In 1910 Curtis Fletcher Marbut[1] delivered a lecture at the University of Missouri before he left there for Washington, DC, to embark on a new career in the U.S. Bureau of Soils (Table 1-1). In this lecture, he stated:

> I have...discussed some of the principles that have already been worked out and some of the problems that are yet unsolved. There is an abundance of work. The laborers in the field are of many degrees. Most of us are mere brickmakers. We labor at the individual units; we have our eyes on the ground and are so intent on our own brick, our own fact, that we fail to see, nay we are often not even interested in, what our neighbor is doing or what relation his brick will have to ours. We each and all bring our small unit of fact and dump it into the same heap with those of our neighbors. We do not see the possibilities that lie in them as units of construction. It is only the rare one among us, the great architect of human generalizations, who is able to bring order out of our chaos and with our units construct the great structures of human thought which we call principles (Marbut, 1913, p. 146).

This chapter is written to recognize both the brickmakers and the architects of the scientific developments leading to the landmark work, *Factors of Soil Formation*, by Hans Jenny (1941). These developments necessarily involve consideration of the history of the discipline called pedology, the study of the origin, nature, classification and distribution of soils. In a broader sense the history reveals the construction of a series of paradigms which range in scope from fundamental factors comprising the universe and the soil to, finally, those factors important in the formation of the soil.

ANCIENT VIEWPOINTS

In the West, the organized understanding of the physical world began with the Greek philosophers: Thales of Miletus reduced nature to one element, water; Anaximenes of Miletus (contemporary of Thales) claimed that the one element was air; and Haraclitus of Ephesus proposed fire as the basic element which when combined with air and water was changed to earth. It was Empedocles of Agrigentum who proposed four elements as the basis for all existence: earth, air, fire, and water. The four elements of Empedocles, developed and refined by Aristotle, became the accepted "Classical model" for almost 2000 yr (Browne, 1944).

Among the Romans, Pliny the Elder recognized that, within the Roman Empire, there were differences among the soils due to color, texture, and vegetation type, and certain crops did better on particular soils than on others (Browne, 1944). Pliny traced the differences among soils to the variable balance within them of the four basic elements of earth, air, fire and water.

The Classical model of four fundamental elements was the basis for understanding the physical world from the eighth through the fourteenth centuries, when knowledge centers were within monastery walls and courts of royalty. The rediscovery of the ancient Greek and Roman philosophers by

---

[1] Birth and death dates of personages discussed in this chapter are given, when known, in Table 1-1.

INTELLECTUAL BACKGROUND OF SOIL FORMATION 3

Table 1-1. Biographical statements on personages discussed in the text.

It is difficult to attach a scientific affiliation to personages since some worked not only in many different areas of a science but in other sciences. We have tried to identify the affiliation of a person which would be relevant to the history discussed in this paper, realizing that a particular individual may be better known for other reasons.

Agassiz, Jean Louis Rodolphe (1807-1873), Swiss and American geologist
Anaximenes of Miletus (fl. ca. 545 BCE), Greek philosopher
Aristotle (384-322 BCE), Greek philosopher
Aubrey, John (1626-1697), British naturalist
Becher, Johann Joachim (1635-1682), German agricultural chemist
Berzelius, Jons Jakob (1779-1848), Swedish agricultural chemist
Black, Joseph (1728-1799), Scottich agricultural chemist
Bogdanov, M., Russian zoologist
Boussingault, Jean-Baptiste (1802-1887), French agricultural chemist
Boyle, Robert (1627-1691), British agricultural chemist
Brewer, William H. (1828-1910), American agricultural chemist
Brongniart, Alexandre (1770-1847), French agricultural geologist
Butlerov, Alexandr Mikhailovich (1828-1886), Russian agricultural chemist
Caldwell, George Chapman (1834-1907), American agricultural chemist
Chamberlin, Thomas Chrowder (1843-1928), American agricultural geologist
Chaptal, Jean Antoine Claude (1756-1832), French agricultural chemist
Chaslavskii, Vasili I. (1834-1878), Russian agricultural chemist
Cook, George Hammel (1818-1889), American agricultural chemist
Cuvier, Georges (1769-1832), French agricultural geologist
Davis, William Morris (1850-1932), American agricultural geologist and geographer
Davy, Sir Humphrey (1778-1829), British agricultural chemist
Dokuchaev, Vasili Vasilievich (1846-1903), Russian agricultural geologist and pedologist
Eaton, Amos (1776-1842), American agricultural geologist
Einhof, Heinrich (?-1808), German agricultural chemist
Emmons, Ebenezer (1799-1863), American agricultural geologist
Empedocles of Agrigentum (ca. 495-535 BCE), Greek philosopher
Fallou, Friedrich Albert (1794-1877), German agricultural geologist and pedologist
Georgi, Johann, Russian agricultural geologist
Glinka, Konstantin Dimitrievich (1867-1927), Russian pedologist
Gyllenborg, Count Gustavus Adolphus, Swedish agricultural chemist and politician
Hilgard, Eugene W. (1833-1916), American agricultural chemist, geologist, and pedologist
Hitchcock, Edward (1793-1864), American agricultural geologist
Horsford, Eben Norton (1818-1893), American agricultural chemist
Humboldt, Alexander von (1769-1859), German geographer and cartographer
Ilienkov, Paval, Russian agricultural chemist
Inostranzev, Alexander, Russian mineralogist
Jameson, Robert (1774-1854), Scottish geologist
Jenny, Hans (1899-1992), Swiss and American agricultural chemist and pedologist
Johnson, Samuel W. (1830-1909), American agricultural chemist
Khodnev, Alexandr, Russian agricultural chemist and physicist
Lavoisier, Antoine Laurent (1743-1794), French agricultural chemist
Leverett, Frank (1859-1943), American agricultural geologist
Liebig, Justus von (1803-1873), German agricultural chemist
Lister, Martin (1639-1712), British agricultural mineralogist and zoologist
Lomonosov, Mikhail Vasilievich (1711-1765), Russian naturalist
Maclure, William (1763-1840), Scottish and American agricultural geologist
Marbut, Curtis Fletcher (1863-1935), American agricultural geologist and pedologist
McGee, WJ (1853-1912), American agricultural geologist
Mendeleev, Dmitri (1834-1907), Russian agricultural chemist
Mitscherlich, Eilhard (1794-1863), German agricultural chemist
Mulder, Gerardus Johannes (1802-1880), German agricultural chemist
Muller, Pieter Erasmus (1840-1926), Danish pedologist

(continued on next page)

Table 1-1. Continued.

Murchison, Roderick Impey (1792-1871), British geologist
Norton, John Pitkin (1822-1852), American agricultural chemist
Orth, Albert (1835-1915), German agricultural geologist and pedologist
Owen, David Dale (1807-1860), American agricultural geologist
Packe, Christopher (1686-1749), British naturalist and agricultural mineralogist
Pallas, Peter Simon (1741-1811), Russian geologist
Paracelsus, Philippus Theoprastus (1493-1541), German agricultural chemist
Pliney the Elder, actually Caius Pinius Secundus (23-79 ACE), Roman natural historian
Ramann, Emil (1851-1926), German agricultural chemist, forester and pedologist
Ritter, Carl (1779-1859), German geographer and cartographer
Rose, Heinrich (1795-1864), German agricultural chemist
Ruprecht, Franz J., Russian botanist
Schubler, Gustav S. (1787-1835), German agricultural chemist
Shaler, Nathaniel Southgate (1841-1906), American agricultural geologist and geographer
Sibirtsev, Nikolai Mihailovich (1860-1899), Russian pedologist
Siliman, Benjamin (1779-1864), American agricultural chemist and geologist
Sovetov, Alexander, Russian agronomist
Sprengel, Carl S. (1787-1859), German agricultural chemist
Stahl, Georg Ernst (1660-1734), German agricultural chemist
Storch, Heinrich Friedrich (1766-1835), Russian agricultural economist
Stremme, Hermann (1879-1961), German agricultural chemist and pedologist
Stuckeley, William (1687-1759), British naturalist
Thaer, Albrecht Daniel (1752-1828), German agricultural chemist
Thales of Miletus (ca. 640-546 BCE), Greek philosopher
Voskresenskii, Alexandr, Russian agricultural chemist
Wallerius, Johann Gottschalk (1709-1785), Swedish agricultural chemist
Werner, Abraham Gottlob (1749-1817), German agricultural mineralogist
Whitney, Milton (1860-1927), American agricultural chemist
Wiegner, Georg (1883-1936), Swiss agricultural chemist
Winchell, Alexander (1824-1891), American agricultural geologist
Young, Arthur (1741-1820), British agriculturalist
Zimin, Nikolai Nikolaievich (1812-1880), Russian agricultural chemist

Renaissance scholars from newly established universities and elsewhere, revitalized scientific inquisitiveness and allowed progress in knowledge to move forward in the fifteenth and sixteenth centuries. One such scholar was Philippus Theophrastus Paracelsus, the Bombast of Hohenheim. He used analytical procedures to determine the proportions in matter of S, Hg, and salt, which he believed comprised the Classical four elements. Paracelsus thought that the soil was uniform in nature and contained these three components or "principles" in the following forms: organic constituents (S), water (Hg), and mineral matter (salt) (Browne, 1944).

After the Renaissance, the fields known today as chemistry, geography, and geology evolved. The scope of these sciences was broad in the early stages of their development, an example being classical mineralogy from which geology and geography emerged (Laudan, 1987). Chemistry and geology were accepted as sciences by the seventeenth and nineteenth centuries, respectively. The agricultural interest within chemistry and geology developed more or less simultaneously within the main stem of each science, since agricultural application was one of the areas of study. In the nineteenth century the "germ" of pedology as we know it today appears to have originated from the interdisciplinary interest of a few natural scientists and the tendency of them and their students to move about for opportunistic or academic reasons. Thus pedology grew out of a complex relation of interdisciplinary and international connections.

AGRICULTURAL CHEMISTRY

The chemical aspects of soils came into focus before any clear geological relationship was documented. The Royal Society of London, particularly cofounder and member Robert Boyle, recognized the agricultural value of the knowledge of soil properties and sent a questionnaire to English farmers concerning the types and conditions of soils on their lands (Georgical Committee, 1665). Northern European chemists also were analyzing and classifying soils in the seventeenth and eighteenth centuries; prominent among these were Johann Joachim Becher and his student Georg Ernst Stahl, and Johann Gottschalk Wallerius and his student Count Gustavus Adolphus Gyllenborg (Browne, 1944). By the late eighteenth century, Scottish chemist Joseph Black had analyzed and classified calcareous marls.

With the rise in France of the new experimental chemistry of Antoine Laurent Lavoisier in the late eighteenth and early nineteenth centuries, scientists in many countries became involved in rigorous soil analyses. The focus of these analyses was the organic matter component and its constituents which for them was the soil. Some of these individuals were: Jean Antoine Claude Chaptal and Jean-Baptiste Boussingault in France; Humphry Davy in England; and Albrecht Daniel Thaer, Heinrich Einhof and their student Carl S. Sprengel (Fig. 1-1), Jons Jakob Berzelius and his students Heinrich Rose, Eilhard Mitscherlich and Gerardus Johannes Mulder, and Gustav S. Schubler in northern Europe (Browne, 1944).

Scientists in northern Europe were particularly active. The term Bodenkunde, i.e., soil knowledge, was first used by Sprengel in his 1837 text (Sprengel, 1837). At the same time Justus von Liebig (Fig. 1-2) and his colleagues and students initiated an academic tradition or school which continued to the early twentieth century and trained many agricultural chemists. Many of these scientists became outstanding teachers: Pieter Erasmus Muller, Emil Ramann, Hermann Stremme, and Georg Wiegner (Fig. 1-3) and his student Hans Jenny in Europe; Alexandr Khodnev, Paval Ilienkov, Alexandr Voskresenskii and his student Dmitri Mendeleev, and Nikolai Nikolaievich Zinin and his student Alexandr Butlerov in Russia; and Eben Norton Horsford, Eugene Woldemar Hilgard, John Pitkin Norton, William H. Brewer, Samuel W. Johnson, and George Chapman Caldwell in the USA (Browne, 1944; Rossiter, 1975; Vucinich, 1970).

AGRICULTURAL MINERALOGY AND GEOLOGY

Simultaneously with the agricultural chemical work, a classical mineralogic knowledge of soils was accumulating, expressed within what today would be called geology, geography and cartography. Royal Society of London

Fig. 1-1. Carl S. Sprengel from portrait in Albrecht Thaer Library, Germany.

Fig. 1-2. Justus von Liebig (from Volhard, 1909).

Fig. 1-3. Georg Wiegner (Courtesy of C. Edmund Marshall family).

members Martin Lister (1684), John Aubrey (1685), and William Stukeley (1724) proposed that soils be analyzed and mapped. The work by Lister (1684) included what appears to be the first soil classification scheme. Later work by Christopher Packe (1743) included a physiographic and soils map of east Kent, England. Soil mapping of the late eighteenth and early nineteenth centuries in Great Britain and Ireland culminated in a series of county-based agricultural surveys made for the British Board of Agriculture—for instance the "General View of Agriculture of the County of Suffolk" by Arthur Young (Young, 1794).

By the end of the eighteenth century, a renewed interest in soils developed from a redefinition of classical mineralogy by German classical mineralogist Abraham Gottlob Werner (Fig. 1-4) into Geognosie, Oryctognosie (modern mineralogy) and Mineral Geography (Ospovat, 1971; Laudan, 1987). Geognosie (in English translated as "geognosy" and eventually replaced with the modern term "geology") was defined as "the abstract systematic knowledge of the solid earth" (Ospovat, 1971, p. 101). That part of geognosy regarding the knowledge of the earth's surface, and as applied to agriculture, was spread through Werner's students. In addition, Werner student Alexander von Humboldt and his student and colleague Carl Ritter spread new geographic and cartographic methods and techniques through Europe and Russia. They influenced their academic colleagues and successors and government officials to adopt these new methods and ideas. Some of the inspired were: Friedrich Albert Fallou (Fig. 1-5), who coined the name Pedologie in 1862 (Fallou, 1862), and Albert Orth in northern Europe; Alexandre Brongniart, Georges Cuvier and his Swiss student Jean Louis Rodolphe Agassiz in France; and members of the Imperial Free Economic Society of St. Petersburg in Russia.

In Scotland, Robert Jameson's interpretation of that part of geognosy concerned with the earth's surface influenced William Maclure. Maclure brought this knowledge, much of which would become known as agricultur-

INTELLECTUAL BACKGROUND OF SOIL FORMATION

Fig. 1-4. Abraham Gottlob Werner (from Ospovat, 1971).

Fig. 1-5. Friedrich Albert Fallou (from Yarilov, 1927).

al geology, to Benjamin Silliman and his students and academic successors in the USA, some of whom were Amos Eaton, Edward Hitchcock, Ebenezer Emmons, David Dale Owen, George Hammel Cook, Nathaniel Southgate Shaler, William Morris Davis, Alexander Winchell, Eugene Woldemar Hilgard, Thomas Chrowder Chamberlin, WJ McGee, Frank Leverett, and C.F. Marbut. The agricultural geologists' developing concept of soil included a recognition that soil was a geologic phenomena separate from others; a rock formation in its own right which was deserving of study (Emmons & Prime, 1845). This discipline became concerned with processes leading to soil formation, such as the weathering of rocks and minerals.

Practicing agricultural geologists formed an international society and began meeting in 1909. The state of the science was summarized by van Baren (1921) in an introduction to a comprehensive bibliographic work on the subject by Wulff (1921). The agricultural geology, or agrogeology, period lasted until the 1920s when the name of the international organization representing that discipline was changed to the International Society of Soil Science.

DEVELOPMENT OF PEDOLOGY

It was Fallou who proposed to elevate the Bodenkunde of Sprengel to an independent science, Naturwissenschaft Bodenkunde or Pedologie. Fallou defined Pedologie as:

> The sum of the knowledge derived by examining and investigating various soils according to one main concept, and put it in a systematic order so that is can be understood; therefore, it is not an economic encyclopedia, or a pre-established set of methods, or an economic science, or an instructional guidebook to apply the results in real life in agriculture (translated from Fallou, 1862, p. 10).

Fallou recognized that a natural-scientific description of soils was not known to the practicing agriculturalists. He also recognized that an understanding of soil distribution and landscape relationships was important. He knew it was important to study and describe the soil both as an undivided whole and as parts of a whole (Fallou, 1862). Fallou claimed that, within Pedologie, soils needed to be studied from many different points of view: geologic, physical, mineralogic, geographic, chemical, and botanical (Fallou, 1862). Fallou believed that pedology must necessarily be an interdisciplinary science.

In Russia, pochvovedenie (translated as both pedology and soil science) developed initially from eighteenth and nineteenth century economic-geographic, chemical, and geologic roots. Early workers included Mikhail Vasilievich Lomonosov, Peter Simon Pallas, Heinrich F. Storch, Johann Georgi, Roderick I. Murchison, Franz J. Ruprecht, and Vasili I. Chaslavskii (Dokuchaev, 1879b). The impetus for soil studies, according to Vucinich (1970), was the Imperial Free Society of St. Petersburg (IFES), whose mission was to evaluate land and improve agriculture. Soil maps had been produced through the IFES since 1838 (Dokuchaev, 1879b). In 1876, the IFES, stimulated by the eminent chemist Dmitri Mendeleev (Fig. 1-6), formed an interdisciplinary commission to study the chernoziom (Vucinich, 1970). The commission members were: Vasili Vasilievich Dokuchaev (Fig. 1-7), geologist-geographer and chair; Mendeleev, Butlerov, and Ilienkov, chemists; Alexander Sovetov, agronomist; Khodnev, chemist and physicist; M. Bogdanov, zoologist; and Alexander Inostranzev, geologist (Dokuchaev, 1879a). Studies were done throughout Russia under the financial sponsorship of the IFES and the St. Petersburg Society of Naturalists in cooperation with the Imperial University of St. Petersburg (Dokuchaev, 1879a,b, 1893; Vucinich, 1970).

It was highly unlikely that scientists would ask questions about the origins of soil until they thought of the object of their study as having an existence of its own. Fallou was moving in this direction, but our reading of his work has so far not identified a clearly articulated set of factors which he thought defined the origin and development of soil. It seems that the Russian investigators—within the interdisciplinary framework of IFES—were the first Europeans to arrive at a view of their subject sufficiently independent to ask such basic questions as what external factors might control the development of soil.

The fundamentals of soil investigation were presented by Dokuchaev in a series of lectures delivered before the IFES in 1877 and 1878 and published in 1879 (Dokuchaev, 1879a). In the first lecture on 24 Feb. 1877, Dokuchaev acknowledged Fallou and Orth as among those who influenced his work. The extent of their influence is uncertain. Perhaps Dokuchaev realized that an important part of Fallou's definition of pedology was the necessity to study the soil profile as a concept or unit of study and certain processes or factors important in soil formation and differentiation. The results of the soil investigations undertaken for the IFES were published in the classic *Russian Chernoziom* (Dokuchaev, 1883).

Fig. 1-6. Dmitri Mendeleev (from personal collection of John Tandarich).

Fig. 1-7. Vasili Vasilievich Dokuchaev (from personal collection of John Tandarich).

Fig. 1-8. Nikolai Mihailovich Sibirtsev (from personal collection of John Tandarich).

Fig. 1-9. Konstantin Dimitrievich Glinka (from Krusekopf, 1942).

The "Dokuchaev school" included such illustrious students as Nikolai Mihailovich Sibirtsev (Fig. 1-8) and Konstantin Dimitrievich Glinka (Fig. 1-9). Under the leadership of Dokuchaev and his students, soil survey investigations began in 1882 in cooperation with the local governmental units, the zemstvos (Vucinich, 1970). Soil classification was developed as a means for comparing soil profiles.

American scientists and laymen first had a glimpse of the "Russian or Dokuchaev School" of pedology in 1893 at the World's Columbian Exposition in Chicago. Dokuchaev and Sibirtsev prepared pamphlets which were translated into English and published in Russia for distribution in Chicago (Dokuchaev, 1893; Dokuchaev & Sibirtsev, 1893). These works contain the main ideas of Russian pedological thought regarding soil genesis and clas-

Fig. 1-10. Hermann Stremme (courtesy of Prof. Dr. Helmut Stremme).

Fig. 1-11. Curtis Fletcher Marbut (courtesy of Western Historical Manuscripts Collection, Archive of Soil Science, University of Missouri—Columbia).

sification to that time. Dokuchaev's works had little initial impact outside of Russia. And for some reason, the contact between American and Russian scientists did not lead to much discourse afterward (Simonson, 1986, 1989; Tandarich et al., 1988, 1990).

Significant impact of the Russian ideas in the USA took place a quarter century after the Columbian Exposition. In the meantime the German and Russian schools of pedology were brought together through the collaboration of Hermann Stremme (Fig. 1-10), a student of Orth, and Glinka, a student of Dokuchaev. Hermann Stremme (Helmut Stremme, 1988, personal communication) helped Glinka write the book *Die Typen der Bodenbildung* (Glinka, 1914). By 1917 Shaler's student C.F. Marbut (Fig. 1-11) had already translated Glinka's work and was applying it to American soils studies. The soil profile concept and the soil-forming factors were widely introduced to American soil scientists through Marbut's translation (Marbut, 1927) and through the First International Congress of Soil Science in 1927 at Washington, DC. Elaboration of the five soil-forming factors was made later by Jenny (1941).

At the same time as Dokuchaev started his work in Russia, German-American Hilgard (Fig. 1-12), a trained agricultural chemist, became well known as an agricultural geologist in the USA. In fact, he was a pioneer pedologist. Hilgard did pedological work in Mississippi and California, although he did not refer to it as such (Hilgard, 1860; Jenny, 1961). He prepared some of the first state and regional soil maps while working for the Bureau of the Census (Hilgard, 1884). He also developed his own profile concept, which was published in 1906 (Hilgard, 1906).

It appears that Hilgard's soil profile concept was used by the Bureau of Soils despite a personal conflict between Hilgard and Milton Whitney (Fig. 1-13), the chief of the Bureau of Soils in the USDA (Jenny, 1961). Hilgard's

INTELLECTUAL BACKGROUND OF SOIL FORMATION

Fig. 1-12. Eugene Woldemar Hilgard (from Jenny, 1961).

Fig. 1-13. Milton Whitney (from Executive Committee of the American Organizing Committee, 1928).

publications were not made available to the soil survey field parties and any use of them was discouraged by Whitney (Jenny, 1961). Consequently, Hilgard was largely unknown to the scientific community outside of California until the 1920s.

In the USA, the interdisciplinary foundations of pedology—chemistry, geology, geography and biology—were not institutionalized within the soil survey program of the USDA, as they had been in Russia by the IFES. The name soil science was used rather than pedology by the USDA. Nevertheless, through the efforts of pioneering pedologists, particularly Fallou, Orth, Hilgard, Dokuchaev, Glinka, Stremme, Marbut and Jenny, the interdisciplinary nature of the science became to be expressed within the model or paradigm known as the factors of soil formation.

SUMMARY

Questions addressed by modern soil science were first asked during the seventeenth and eighteenth centuries by researchers in other disciplines. During the late nineteenth and early twentieth centuries, that branch of soil science called pedology developed as an independent discipline from the traditions of agricultural chemistry, geography, and geology in Europe, Russia, and the USA. Disciplinary boundaries were very diffuse in the nineteenth century and it was relatively easy to work in several of today's narrowly defined specialties. The development and spread of the factors of soil formaton paradigm in pedology came from a complex evolution of interdisciplinary and international connections. Pioneer pedologists, such as Fallou, Hilgard, Dokuchaev, Glinka, Orth, Stremme, Marbut and Jenny, considered the science to be interdisciplinary from the outset. The factors of soil formation arose from pedology's interdisciplinary origins.

ACKNOWLEDGMENTS

The authors wish to thank the following for their assistance with the translation of original materials: Dr. Heinz Damberger and Dr. Leon R. Follmer, Illinois State Geological Survey, Champaign, and the late Dr. Dimitri Shimkin, Department of Anthropology, University of Illinois at Urbana-Champaign. Background research for this paper was done by the senior author while a graduate student at the University of Illinois at Urbana-Champaign under the supervision of the following whose assistance is gratefully acknowledged: Dr. Robert G. Darmody, Dr. Thomas J. Bicki, and the late Dr. Ivan J. Jansen, Department of Agronomy; Dr. Leon R. Follmer, Illinois State Geological Survey; Dr. Donald L. Johnson, Department of Geography; Dr. W. Hilton Johnson, Department of Geology; and Dr. Evan M. Melhado, Departments of History and Chemistry.

REFERENCES

Aubrey, J. 1685. Memoires of natural remarques in the county of Wilts. London.

Browne, C.A. 1944. A source book of agricultural chemistry. Chron. Bot. 8(1):1-290.

Dokuchaev, V.V. 1879a. Tchernozeme (terre noire) de la Russie D'Europe. Soc. Imperiale Libre Econ. Imprimeric Trenke & Fusnot, St. Petersburg, Russia.

Dokuchaev, V.V. 1879b. Cartography of Russian soils. (In Russian.) Ministerstva Finansov, St. Petersburg, Russia.

Dokuchaev, V.V. 1883. Russian chernoziom (In Russian). St. Petersburg, Russia.

Dokuchaev, V.V. 1893. The Russian steppes. Study of the soil in Russia in the past and present. Dep. of Agric. Ministry of Crown Domains for the World's Columbian Exposition at Chicago, St. Petersburg, Russia.

Dokuchaev, V.V., and N.M. Sibirtsev. 1893. Short scientific review of Professor Dokuchaev's and his pupil's collection of soils exposed in Chicago in the year 1893. St. Petersburg, Russia.

Emmons, E., and A.J. Prime. 1845. Agricultural geology. Am. Q. J. Agric. Sci. 2(1):1-13.

Executive Committee of the American Organizing Committee. 1928. Proc. Pap. Int. Congr. Soil Sci. 1st, Washington, DC. 13-22 June 1927. Vol. I. Am. Organ. Comm. Int. Congr. Soil Sci., Washington, DC.

Fallou, F.A. 1862. Pedologie oder allgemeine und besondere Bodenkunde. Verlag Schonfeld, Dresden, Germany.

Georgical Committee. 1665. Enquiries concerning agriculture. Philos. Trans. R. Soc. London 1:91-94.

Glinka, K.D. 1914. Die Typen der Bodenbildung. Verlag von Gebruder Borntraeger, Berlin.

Hilgard, E.W. 1860. Report on the geology and agriculture of the state of Mississippi. E. Barksdale, State Printer, Jackson, MS.

Hilgard, E.W. 1884. Report on cotton production in the United States. Vol. 1 & 2. Census Office, Dep. of the Interior. U.S. Gov. Print. Office, Washington, DC.

Hilgard, E.W. 1906. Soils. The Macmillan Co., New York.

Jenny, H. 1941. Factors of soil formation. McGraw-Hill Book Co., New York.

Jenny, H. 1961. E.W. Hilgard and the birth of modern soil science. Collana Della Rivista Agrochimica no. 3. Pisa, Italy.

Krusekopf, H.H. (ed.). 1942. Life and work of C.F. Marbut. SSSA, Madison, WI.

Laudan, R. 1987. From mineralogy to geology: The foundations of a science, 1650-1830. Univ. Chicago Press, Chicago and London.

Lister, M. 1684. An ingenious proposal for a new sort of maps of countrys. Philos. Trans. R. Soc. London 14:739-746.

Marbut, C.F. 1913. Geology. Univ. Missouri Sci. Ser. Bull. 1:125-146.

Marbut, C.F. (trans.) 1927. The great soil groups of the world and their development by K.D. Glinka. Edwards Brothers, Ann Arbor, MI.

Ospovat, A.M. (transl.) 1971. Short classification and description of the various rocks by Abraham Gottlob Werner. Hafner Publ. Co., New York.

Packe, C. 1743. Ancographia; sive, convallium descripto. J. Abre, Canterbury, England.

Rossiter, M.W. 1975. The emergence of agricultural science/Justus Liebig and the Americans, 1840-1880. Yale Univ. Press, New Haven and London.

Simonson, R.W. 1986. Historical aspects of soil survey and soil classification. Part I. 1899-1910. Soil Surv. Horiz. 27(1):3-11.

Simonson, R.W. 1989. Soil science at the World's Columbian Exposition, 1893. Soil Surv. Horiz. 30(2):41-42.

Sprengel, C.S. 1837. Die Bodenkunde oder die Lehre vom Boden nebst einer vollstandigen Anleitung zur Chemischen Analyse der Ackereden. I. Muller, Liepzig, Germany.

Stuckeley, W. 1724. Itinerarium curiosum. London.

Tandarich, J.P., R.G. Darmody, and L.R. Follmer. 1988. The development of pedologic thought: Some people involved. Phys. Geogr. 9:162-174.

Tandarich, J.P., R.G. Darmody, and L.R. Follmer. 1990. Some international and interdisciplinary connections in the development of pedology. p. V191-V196. *In* Trans. Int. Congr. Soil Sci., 14th. 1990, 5, 5. Int. Soc. Soil Sci., Kyoto, Japan.

van Baren, J. 1921. Agrogeology as a science. p. 5-9. *In* A. Wulff (ed.) Bibliographia agrogeologia. Mededeelingen van de Landbouwhoogeschool Deel 20. H. Veenman, Wageningen, The Netherlands.

Volhard, J. 1909. Justus von Liebig. Verlag von Johann Ambrosius Barth, Leipzig, Germany.

Vucinich, A. 1970. Science in Russian culture 1861-1917. Stanford Univ. Press, Stanford, CA.

Wulff, A. 1921. Bibliographia agrogeologia. Mededeelingen van de Landbouwhoogeschool Deel 20. H. Veenman, Wageningen, The Netherlands.

Yarilov, A.A. 1927. The Russian pedologists work in the sphere of the history of their science. Pedology 22(2):5-20.

Young, A. 1794. General view of the agriculture of the county of Suffolk. C. Macrae, London.

2 Factors of Soil Formation: Contributions to Pedology

L. P. Wilding

Texas A&M University
College Station, Texas

ABSTRACT

Factor's of Soil Formation" was a pedological classic that succinctly synthesized into a conceptual model soil concepts developed earlier by Dokuchaev and associates in Russia. It further articulated methods by which the system could be quantitatively studied in a more scientific approach to pedology. This forward-looking treatise focused mostly on uncultivated soils developed under short geologic time scales from relatively "uniform" glaciated parent materials. The functional relationships developed were constrained by geographical limits and analytical databases. Interdependence of state factors limited extension of models, especially to older landforms. Major contributions were: (i) a better appreciation of the Russian pedological works, (ii) a conceptual framework to comprehend soil distribution patterns, (iii) a methodology for pedological quantification, (iv) a stimulus to develop soil genesis models, (v) a basis for construction and quantification of soil taxonomy, and (vi) a synergistic ecosystem approach that rallied many diverse interests and approaches of pedology into one framework.

The objective of this paper is to evaluate the scientific contributions and merits of Hans Jenny's text *Factors of Soil Formation* (Jenny, 1941). At the outset I would submit that this evaluation is subjective and may not be shared by all professionals. In this evaluation no attempt was made to conduct a comprehensive literature review. Rather an overview is presented of Hans Jenny's early scientific career, the state of pedology in 1941, major contributions of Jenny's work, limitations of the factorial approach to pedology, and other pedogenic models stimulated by his work. In this analysis, major focus has been given to the text *Factors of Soil Formation*, but it has been impossible to cleary discriminate other contributions of Jenny including his impact on students and associates, cutting-edge scientific contributions, interdisciplinary contributions to ecology and services to educational and professional societies and institutions.

Copyright © 1994 Soil Science Society of America, 677 S. Segoe Rd., Madison, WI 53711, USA. *Factors of Soil Formation: A Fiftieth Anniversary Retrospective*. SSSA Special Publication 33.

The approach followed was to survey selected literature with special reference to recent pedological texts, to consider major breakthroughs in pedology during the past 50 yrs, to counsel associates and students of Hans Jenny, and to obtain input from officers and peers of Divisoin S-5, and other professionals who have contributed significantly to the discipline of pedology.

WHAT IS PEDOLOGY?

The term "pedology" is used widely throughout the world but in different contexts. In the European sense, pedology has been used synonymously with "soil science" (Buol et al., 1989, p. 3). In contrast Buol et al. (1989) defined pedology as "a phase of soil science that deals with factors and processes of soil formation including description and interpretation of soil profiles, soil bodies and patterns of soil on earth's surface." The concept of pedology as used in this paper is the science of soil development (Sposito & Reginato, 1992). More specifically pedology is defined herein as *"that component of Earth science that quantifies the factors and processes of soil formation including the quality, extent, distribution, spatial variability and interpretation of soils from microscopic to megascopic scales."*

Pedology is both an interpolative and extrapolative science (Wilding, 1988; Sposito & Reginato, 1992). This discipline provides a hierarchical framework to integrate components of soils (e.g., mineral structures, mineral-organic complexes, soil aggregates, and soil horizons) into basic soil individuals called pedons (Fig. 2–1). Pedons are assembled into toposequences, which in turn comprise regional physiographic units that collectively represent the pedosphere or global soil cover. Pedology provides the basic framework to examine soils at variable scales of resolution using different methodologies so the system can be viewed in holistic terms for component integration and system extrapolation (Fig. 2–2). The framework of pedology is based on Jenny's paradigm of soils as a function of the five soil-forming factors—climate, organisms, parent material, topography and time. Other unspecified factors also were included in this paradigm.

JENNY: THE SCIENTIST, TEACHER AND SCHOLAR

The following comments about Jenny's early career development are based heavily on the forward "Hans Jenny and Fertile Soil" by Jerry Olsen (Jenny, 1980) and Jenny's *Oral History* (Jenny, 1989). Dr. Jenny was born in 1899 in Basil, Switzerland. He died in 1992 in Oakland, CA. He was a naturalist with broad interests in ecology and considered soils as the basic natural resource governing ecological systems. His formal training was at the Swiss Federal Technical Institute, Zurich, where he worked under the tutelage of Professor Georg Wiegner for his Ph.D., which he received in 1927. He was a physical chemist by training with pioneering research on ion exchange related to colloidal chemistry—special focus was on the humus frac-

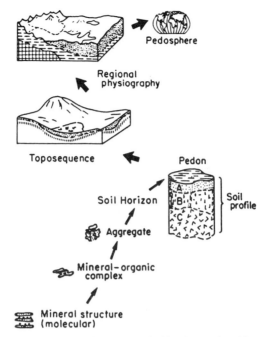

Fig. 2-1. Soil components and systems at various hierarchical levels. (reprinted from *Opportunities in Basic Soil Science Research*, a miscellaneous publication by SSSA, edited by Sposito and Reginato, 1992).

tion. Professor Wiegner encouraged him to become a pedologist and to pursue the work of E.W. Hilgard (1921) and V.V. Dokuchaev (1893). This was the birth of pedology for him. His field experiences were mostly self-guided in his early career while on excursions to the Alps of Switzerland and later in the USA.

He came to the USA in 1927 to work with S.A. Waksman on colloidal chemistry of humus. Later, while at Rutgers University (New Brunswick, NJ) he worked with the New Jersey Agricultural Experiment Station on ion exchange in barley; clearly, his early training and interests were a hybridiza-

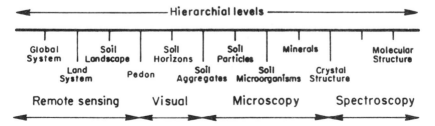

Fig. 2-2. Schematic illustration of the relationship between increasing levels of resolution and field of view, as related to hierarchical levels (reprinted from *Opportunities in Basic Soil Science Research*, a miscellaneous publication by SSSA, edited by Sposito and Reginato, 1992).

tion of pedology, colloidal physical chemistry and organic matter biochemistry.

He attended the 1st International Congress of Soil Science (ICSS) in Washington, DC, in 1927; this Congress and the tour that followed were pinnacle to his early professional career development because at this ICSS, he was introduced to the "giants" of pedology. These contacts continued to motivate his incessant curiousity about the "whys of nature." At the ICSS he met C.F. Marbut, who Jenny considered the greatest soil scientist of the era. Marbut likewise was impressed with Jenny and considered his Ph.D. research on ion exchange as the best paper presented at the Congress. During the post-Congress transcontinental tour of the Red soils (Ultisols) in southern the USA, and Black soils (Mollisols) of Canada, Jenny theorized that the C–N ratios and organic matter contents of soils may be more influenced by the temperature than by precipitation. He believed it was possible to better partition these climatic variables in North America than had been possible in previous Russian work because of the non-synchronous precipitation–temperature patterns in North America. Further, while on the post-Congress tour, he met R. Bradfield, Soils Department, University of Missouri, who invited him to join the staff as an instructor.

Hans Jenny joined the Department of Soils, University of Missouri, in 1927 and remained there until 1936. He rapidly climbed through the professional ranks to professor. This was an enriching experience for him with M.F. Miller's lectures on soil classification, pedological field trips with H.H. Krusekopf in the loesial prairies and forested Ozarks, and interactions with W.A. Albrecht, a microbiologist investigating soybean responses conditioned by soil texture and cation exchange. Jenny considered higher education in the USA system much more invigorating than the more prestigious European systems because of the breadth of subject matter areas represented, integrated scientific expertise, team orientation, and less domineering faculty.

During his tenure at the University of Missouri, he analyzed a wide array of soils across the Great Plains, as a function of climatic gradients and compared prairie and forested conditions with cultivated counterparts. Within a decade he had quantified N time curves, N climatic trends, losses of organic carbon (OC) and N from cultivated soils, and developed simple mathematical models for N and C balances. By his own admission, he had "rediscovered" the five soil forming factors of E.W. Hilgard and V.V. Dokuchaev, but he would argue the N-climatic three-dimensional surface relationships were truly an original contribution to pedology (Jenny, 1980). He enjoyed formulating numerical models of field-observed morphological properties and gained considerable respect and recognition at professional society meetings for these stimulating contributions.

In 1936 he accepted an invitation to join the faculty at the University of California, Berkeley. He taught courses in pedology and colloidal chemistry and later assumed E.W. Hilgard's courses and responsibilities in maintaining Berkeley's reputation and commitment to pedology. It is here where he developed a more comprehensive extension of the factor scheme and numerical relationships to pedogenic property functions. During his early tenure at Berkeley he struggled with definitions and causes of soil variability. Why did physical chemists choose temperature and pressure as critical variables? What puts time and climate together with soils? How did parent material fit into the scheme? It was through these challenges and carefully designed experiments with natural soil systems that he capitalized on the state factor model to elucidate relationships among soils properties previously unknown.

It is clear that the attributes that Jenny possessed as a scientist were as a peer, teacher, mentor, intellect, physical chemist, pedologist, naturalist and philosopher. In addition to his text *Factors of Soil Formation: A System of Quantitative Pedology* (Jenny, 1941), he published nine other prestigious books and monographs, over 150 refereed scientific technical articles and presented numerous invitational lectures. Likewise, he received a number of prestigious awards including several honorary doctorates, society fellowships and life memberships to professional organizations. He was a leader of the scientific community, an active soils environmentalist and a strong preservationist. Jenny's models of soil as a macrosystem that culminated in the "state factor equation," and as a microsystem that focused on resolution of the root-soil boundary interfaces, embrace modern concepts of pedology (Sposito & Reginato, 1992). These scientific contributions demonstrated Jenny's leadership in soil science well ahead of his time. Division S-5, Soil Science Society of America, commemorated the golden anniversary of the publication of his text (Jenny, 1941) by holding a symposium in his honor in 1991. It is in this spirit that this paper is written.

SETTING OF PEDOLOGY IN 1941

Tandarich (1991) and Arnold (1991) have already outlined historical perspectives of this period. Cline (1961) also outlined the status of pedology from it's early roots that covered this period. Several aspects are clear. Soil-forming factors as *soil formers* were recognized by Dokuchaev (1893) and associates long before 1941. The factorial functional model was championed in the USA by E.W. Hilgard, C.F. Shaw, G.N. Coffey and later by C.E. Kellogg and Jenny (Arnold, 1983, 1991; Jenny, 1980). The generalized process model (podzolization, laterization, calcification, etc.) also was in vogue at this time (Byers et al., 1938). Pedologists of that period were attempting to distance themselves from their geological parentage much in the same way as some pedologists today may wish to distance themselves from our parentage of factorial analysis.

It was at the close of the Dust Bowl Era (1934–1938) and the beginning of the New Deal Era by President F.D. Roosevelt that a wave of soil appreciation was sparked by the erosion across the Great Plains. Soil became a national concern and soil conservation was taught in schools. The Soil Conservation Society was born and the soil survey program was under its jurisdiction for utilitarian purposes. The major focus was to inventory the national soil resources for soil erosion control and development of farm plans.

This era represented the period of World War II, with the declaration of the war by the USA in the same year that Jenny's book was published. The brightest young minds of pedology were inducted into the armed forces to serve their country. The World Geography Group, Soil Survey Division, USDA-SCS was assembled to help in this behalf. Middle and senior career pedologists were employed in teaching and research appointments at universities and agricultural experiment stations, and with state/federal agencies to conduct land resource inventories. Most senior-level pedologists currently active today were either not yet born, or in their infancy or childhood.

The most productive, dynamic and prestigious pedological research and teaching programs were in the north central and northeastern sectors of the USA where soils had developed primarily from glaciated materials. Here, the pedogenic time frame was relatively short and many soil conditions were less well recognized, except for the Unviersity of California at Berkeley, which enjoyed a long-established pedology program of excellence under the tutelage of Hilgard, Shaw and Jenny.

Compared with today's voluminous state/federal soil characterization bases, the physical, chemical, mineralogical and biological information available at that time was seriously constrained; many analyses were of total silicates or oxides. Likewise, information was heavily biased towards uncultivated conditions without much knowledge or appreciation for the magnitude, mode and mechanisms of small-scale or large-scale spatial variability (Wilding & Drees, 1983).

The understanding of soils relative to time stratigraphic units, geomorphic surfaces and absolute chronology was not fully appreciated (Cline, 1961). The pioneering work of Ruhe (1969) and associates in Iowa and later the establishment of the soil/geomorphology projects in New Mexico, North Carolina, and Oregon by the USDA-SCS, under the leadership of Guy D. Smith, had not yet started (Arnold, 1991). The period predates the use and application of computers to establish mathematical models, multivariant analysis, multiple regression relationships or to manage data bases.

Widely used textbooks of the day were *Pedology* (Joffe, 1936), *Soil Survey Manual* (Kellogg, 1937), *Soils and Men* (USDA, 1938), *Properties of Colloids* (Jenny, 1938), *Factors of Soil Formation* (Jenny, 1941), *Nature and Properties of Soils* (Lyon & Buckman, 1938), and *Soil Conditions and Plant Growth* (Russell, 1937). The major outlets in the USA for soils literature were the *American Soil Survey Association Proceedings* (1920–1936), *Soil Science Society of America Proceedings* (1937–1975) and *Soil Science* (1916–present).

CONTRIBUTIONS OF *FACTORS OF SOIL FORMATION*

While the concept of soil-forming factors incorporated into a factorial equation was not original to Jenny (1941, 1980), his pioneering work demonstrated methods by which the soil system could be quantitatively investigated. It engendered a more scientific approach by synthesizing conceptual ideas into a mathematical syllogism—a more scientific grounding to pedology. He properly cited pioneering contributions to Dokuchaev, Hilgard, Shaw and Tuxen (Jenny, 1980, p. xi, 203). Kellogg (1936, p. 9) also had employed the factorial equation in teaching pedology and practicing the discipline. Kellogg was much more involved in the geographical aspects of soil inventory and classification while Jenny was the university academician steeped in pedological quantification endeavors. Both served useful roles in enhancing the image of pedology and wise stewardship of the most pervasive and fundamental of all natural resources.

Smeck et al. (1983) comprehensively evaluated the strengths and weaknesses of pedogenic models including the factorial analysis model of Jenny (1941). The factorial model states that soil (S) is a function of climate (cl), organisms (o), topography (r), parent material (p) and time (t). It is expressed as

$$S = f(cl, o, r, p, t, \ldots)$$

where the dots indicate additional unspecified factors. Accordingly, the factors define the soil in terms of controls on pedogenesis and soil distribution factors—"an environmental formula" (Jenny, 1941, p. 16). They were neither causes nor forces, as claimed by many critics, but factors that define the "state and history of a soil" (Jenny, 1980, p. xi). Jenny believed firmly that the equation could be solved under ideal conditions and that the variables were independent (Jenny, 1980, p. 203), though he recognized they also may be interrelated (Jenny, 1941, p. 16; and Jenny, 1980, p. 203). He believed the independency criterion was widely misunderstood and that he was overly maligned by that criticism. While he stated "The fundamental equation of soil formation (4) is of little value unless it is solved" (Jenny, 1941, p. 7), I do not share this perspective. It has proven to be the most valuable tool in teaching soil genesis, even if it cannot be numerically resolved for many soils. The initial state factor, parent material, included the physical, chemical and mineralogical composition of the inorganic and organic components. The initial state also was conditioned by topography. Vegetation seemed to take precedence over faunal aspects as controls on pedogenesis though he considered both in his works. Anthropogenic effects on soils were considered in his first text (Jenny, 1941, p. 232-257), but later given greater emphasis when a special symbol was introduced for the human organismal component (Jenny, 1980, p. 203). Jenny struggled with the biotic factor as a real dilemma. He knew that vegetation affected soils and soils affected vegetation, but didn't like the circuitous argument. Organisms were not clearly considered in Jenny's 1941 text but later this factor was defined as controlling floristic potential or expression (Jenny, 1961, 1980). It appears that Jenny considered climate as the major pedogenic driving vector acting through time with vegetation, topography and parent material serving as secondary controls.

Jenny (1941) believed the two major contributions of this text to pedology were:

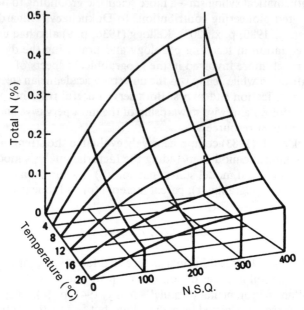

Fig. 2-3. Model of soil N variations as a function of temperature and a moisture index (NSQ) for the Great Plains region of USA from Canada to Mexico (from Jenny, 1980).

1. Redefining the factors as "possible" *independent* variables in the soil-forming factor equation; and
2. Development of the three-dimensional N climate surface net which illustrated clearly for the first time the dependency of soil N contents on temperature and precipitation variables; this may not seem profound today, but consider the data base and computational tools of the era (Fig. 2-3).

He said the model has been presented before but "...I can solve the equation. That was the new approach" (Jenny, 1980, p. xii).

Contemporaneous pedologists would add the following contributions of Jenny's text:

1. Communication of Russian pioneering work.
2. Conceptual framework to clearly and consiely comprehend all distribution patterns.
3. Methodology for pedogenic quantification.
4. Stimulus to develop more comprehensive process-oriented pedogenic models.
5. Basis for construction and quantification of *Soil Taxonomy*.
6. Ecosystem approach to address environmental quality and global climate change.

Many of the pedogenic functions have been summarized in texts and special publications including Jenny (1941, 1980), Wilding et al. (1983a,b),

Ruhe (1975), Yaalon (1971), Buol et al. (1989), Birkeland (1984), Fanning and Fanning (1989), Singer and Munns (1991), and Sposito and Reginato (1992). According to Jenny (1941, p. xi) the ultimate goal of functional analysis was the formulation of quantitative laws that permit mathematical treatment. This permitted quantitative linkage of soil geography and functional analysis. However, no correlation had been found under field conditions between controlling factors and soil properties that "satisfied the requirement of generality and rigidity of natural laws" (Jenny, 1941, p. xii).

Communication and Conceptualization

Jenny (1941) contributed a better understanding and appreciation of the Russian pioneering work in pedology to the Western World, especially the work of Dokuchaev. It stimulated the communication of Soviets with pedologists in the Western World. The text inspired students and teachers. It was widely read, critiqued, and utilized as a reference and classroom text. I dare say that most soil scientists today still conceptualize and verbalize soils into the generalized framework of Jenny's factorial model. The text had an inordinate impact on the teaching and research agenda of pedology for the past 50 yrs. It has heavily influenced modern pedology and geomorphology texts in this country and helped establish order of our understanding of soil geography. It enhanced interdisciplinary communication with ancillary disciplines including ecology, geomorphology, climatology, and geology.

Quantification

The text was pinnacle in bringing pedology to a more quantitative and less subjective/descriptive scientific approach. This added credibility to a discipline suffering from appendage to agriculture, perceived low esteem by others and embracing only a utilitarian focus. Ordination of factors provided a numerical solution for many bio-, litho-, climo-, chrono-, and toposequence functions widely reported in the literature. It provided a means to dissect landscapes into segments along vectors of state factors for better understanding. Twenty-eight percent of the nearly 600 papers published in Division S-5, of the *Soil Science Society of American Proceedings* (*Journal*) have dealt with soil-forming factor analysis (Wilding, 1986). Over 70% of these papers emphasized soil parent material and landform evolution in understanding spatial variability of soils. This reflects evolving emphasis given to geomorphology and stratigraphy as determinant variable sto pedogenesis, soil landscape patterns, chronology, land use, and soil behavior, particularly at large scale (close-interval) resolutions. In summary Jenny's early work was a search for quantitative dimensions.

Construction of Soil Taxonomy

Though not generally recognized nor appreciated, Jenny's text contributed to *Soil Taxonomy* (Soil Survey Staff, 1975). Quantitative limits were placed on classes using both factor controls directly and morphogenetic properties as marker expressions of pedogenic development that was controlled by state factors. His 1941 text discusses alternative approaches to soil classification (Jenny, 1941, p. 263-267). Major factor controls of pedogenesis were used at order level and secondary controls at the suborder level in *Soil Taxonomy*. Dr. Guy Smith, a graduate student of Jenny and "father of Soil Taxonomy," clearly was influenced by Jenny's work to couple state factors and morphological properties in a classification system. Indirectly, Jenny's search for quantitative parameters and dimensions influenced the construction of fixed-class boundaries in soil taxonomy and the rationale for these arbitrary divisions.

Ecological Dimensions

The text is currently contributing to a more comprehensive and broader understanding of environmental issues and challenges. Because this topic will be covered in detail by Vitousek (1991) as a component chapter in this text, only pertinent contributions will be cited:

1. Scaling soil ecosystem observations from the microscopic to megascopic environmemtal levels.
2. Identifying mechanisms of C sequestration and biomass accumulation in soils.
3. Partitioning the impacts of climate and vegetation components on C and N cycles.
4. Quantifying rates of C decomposition and emission of CO_2 to the atmosphere.
5. Determining likely locations of methane production.
6. Establishing interactions among ecological/pedological disciplines.

He also has been an effective spokesperson for preservation of ecosystems and minimizing adverse anthropogenic impacts to the environment (Jenny, 1989).

LIMITATIONS OF STATE FACTOR MODEL

Limitations to be considered herein were generally recognized in Jenny's 1941 or 1980 texts. In many cases, the limitations were constraints of state-of-knowledge of that period. In other cases there are differences in scientific rigor, perspective, professional judgements and intellectual approaches. They are presented herein to serve as a vehicle to stimulate scientific discussion and dialogue. Pedogenic models have been reviewed, critiqued and strengths and weaknesses evaluated by Crocker (1952), Smeck et al. (1983), and Bryant and Olson (1987). The following comments draw partially on these references. Perhaps one of the major limitations of the state factor model as viewed from breakthroughs in modern pedology is the *recognition and general acceptance that soils are developed along polygenetic pathways, on dynamically evolving landforms under the influence of paleoclimates, in non-uniform parent materials and through combinations of processes* (Crocker, 1960; Cline, 1961; Arnold, 1965; Simonson, 1978; Wilding, 1986). This has enhanced the morphogenetic model of soil and moved us farther from purely genetic concepts. It is this backdrop that will serve as a yardstick in evaluating possible constraints of Jenny's factorial model. The following limitations will be briefly considered:

1. Independence of state factors.
2. Extension to older landforms.
3. Polygenetic pathways of soil genesis.
4. Factor interchangeability.
5. Nonsystematic spatial variability.
6. Suitable data base.
7. Anthropogenic influences.
8. Knowledge of precise processes.
9. Difficulty in testing and validating models.

Most pedologists have strong reservations about the rigorous independence of state factors. Climatic records, organismal dynamics, geomorphology, and paleopedology argue against state factor independence (Stephens, 1947; Crocker, 1960; Stevens & Walker, 1970; Yaalon, 1975). Jenny recognized this fact but searched for examples where factor interaction would be minimized. Application of the factorial model to recent or late Quaternary landforms probaby has the best opportunity for success. Extensions to older landforms are more risky because of greater probability of factor interactions and multiple polygenic pathways of soil genesis (Crocker, 1960). Examples of this have been reported by Simonson (1978), Arnold (1965) and Wilding and Flach (1985) for many properties and diagnostic horizons in soils including: albic, spodic, cambic, argillic, calcic, mollic, duripan, and fragipan. Pedogenesis in soils can establish new *internal environmental controls* by changes in texture, acidity, drainage and hydrology and fertility. Because of this, factor changes may exert different *external environmental controls* through time (Simonson, 1978; Runge, 1973; Vepraskas & Wilding, 1983). Further, it is believed that the influence of parent materials (e.g., polycycled pre-weathered sediments) can yield the same or very similar chemical attributes in a soil as another soil subjected to strong in situ weathering because of intensive soil leaching (climate). An example of this is the formation of kandic and oxic horizons in many tropical regions (pedogenic vs. geogenic; Van Wambeke, 1983). Likewise, the effect of parent materials (e.g., infertility or toxicity) can result in a feedback mechanism controlling biomass production, and in turn organic matter accumulation, in a humid environment much the same as climate may control these processes in a more xerophytic one (Smeck & Runge, 1971; Runge, 1973).

Spatial and temporal variability is much greater and at closer intervals than generally recognized. For many soil properties the coefficients of variation are 25% or greater over intervals of a few meters or less (Wilding & Drees, 1983). Both systematic and random variation are of a temporal and static nature (Wilding & Drees, 1983). This variability is generally highest with dynamic variables and least with more permanent properties. This will cause high noise in factorial analysis, an observation that Jenny also recognized (Jenny, 1941, p. 19–20).

Jenny's 1941 text suffered from geographical and geomorphological controls of the data base. There was little clay mineralogy, geomorphology or parent material control for much of the data collected. Jenny noted in his 1941 text that topography had not been given the attention it deserved. This is an area where significant advances have been made since 1941. Jenny may not have been aware when he wrote the text that geomorphic evolution occurs at a time scale that approximates pedogenic change. Furthermore, that "old and young" landforms occur within very close proximity as a function of geomorphic stability. Furthermore, there is little evidence in Jenny's early work that indicates how sampling was conducted and what assurances were taken to satisfy parent material control in populations sampled. If we look closely enough, no soil forms in a single homogeneous parent material. How does this fact confound the numerical relations Jenny derived? How did the parent materials change subtly, progressively or markedly across the broad geographical transects he sampled? I do not propose to answer these questions, but later, Jenny (1980) fully appreciated these limitations.

Difficulty in solving the factorial equation occurred because of the lack of incremental data sets for factor variables which prevented developing pedogenic rate changes (Bryant & Olson, 1987). Changes in properties are generally not linear with time and thus such incremental data were critical to the rate function analysis. Problems also occurred in obtaining partial differentials with nonoverlapping factors or coupled factors (Runge, 1973). The data base that was available was generally insufficient for general solution of the factorial equation in a rigorous manner.

The human influence confounds factor variables—liming, drastically disturbed lands, drainage of wetlands, compaction, irrigation, salinity/sodicity, etc., (Fanning & Fanning, 1989). Anthropogenic influences on pedogenesis were not outside the scope of the first text by Jenny (1941) but were given more prominent consideration in his last text (Jenny, 1980).

The factorial analysis doesn't elucidate pedogenic processes involved in soil genesis so these influences could not be partitioned (Smeck et al., 1983). Likewise, it is not possible to mathematically solve the factorial model in terms of specific processes because rate functions are unknown (Bryant & Olson, 1987). It is difficult to impossible to rigorously reconstruct the time effects on pedogenesis (Yaalon, 1971). Numerous chronosequences have been offered in the literature (e.g., Birkeland, 1984), but establishing such effects on resultant soil attributes are still open to question. An exception may be for those property functions that establish steady-state over short geological time periods (a few hundreds–thousands of years).

Finally the multiple working hypotheses is not fostered in utilizing factorial analysis. Only one dominant controlling variable is attributed to a property without considering alternatives or interactions. This doesn't consider multiple origins of soil properties or combinations of processes. It constrains destruction or revision of models and recycling hypotheses for further model refinement and revision (Cline, 1961). Scientific historians (e.g., Kuhn, 1970) argue that any model, or concept, restricts thought because, by its very nature, it determines the appropriate questions to ask and appropriate data to collect. This is why Chamberlain (1897) and later Platt (1964) argue vigorously that the scientific methods of (i) ruling theory, and (ii) single multiple working hypothesis are hazardous. They contend these intellectual methods, do not promote thoroughness of scientific inquiry, guard against adoption of tentative theory and ward off intellectual affection or parentage. On the contrary, establishment of multiple working hypotheses promotes bringing all rational explanations in view, development of every tenable and testable hypothesis, and exposure of imperfections in our knowledge base. Strong inference (Platt, 1964) invokes execution of crucial experiments to exclude some of the alternative hypothesis and adopt what is left. This is analogous to climbing a logic tree. Full disclosure of cause and effects in complex soil systems is multiple in character, polygenetic in function, and consequences of combinations or sets of factor interactions.

NEW PEDOGENIC MODELS

The functional factorial model that Jenny (1941) popularized stimulated vigorous development of new pedogenic models incorporating all, some or none of the factorial approach. The progression of model development includes the following: the state-factor model (Jenny, 1941), extended state-factor model (Jenny, 1961, 1980), systems mass-balance process model (Simonson, 1959; Chadwick et al., 1990), energy flux model (Smeck & Runge, 1971; Runge, 1973), chemical equilibrium residua and haplosoil model (Chesworth, 1973a,b), soil-landscape systems models (Ruhe, 1969; Huggett, 1975), progressive-regressive evolutionary model (Johnson & Watson-Stegner, 1987; Johnson et al., 1990), coupled reactions–factors–processes model (Ciolkosz et al., 1989), and simulation models (Levine & Ciolkosz, 1986, 1988; Bryant & Olson, 1987; Hoosbeek & Bryant, 1992). Models are essential tools of science. They provide a perspective, a way of organizing thoughts and pondering facts into a conceptual framework. Simplicity of reality is commonly the best model. Models must generate testable hypotheses to separate cause and effect. If not, established theories of science can restrict thought (Chamberlain, 1897; Cline, 1961).

While most models for soil genesis have been conceptual or verbal, more quantitative mathematical simulation and system models are the hope of the future (Dijkerman, 1974; Huggett, 1976; Bryant & Olson, 1987). Gerrard (1981), Smeck et al. (1983), Bryant and Olson (1987), Johnson and Watson-Sieger (1987), and Hoosbeek and Bryant (1992) have recently reviewed many pedogenic models. Even for many simulation models offered today, the factorial approach is utilized as controls that govern direction and magnitude of the specific pedogenic processes being simulated.

Recent attempts have been made to use multivariant statistical analysis to define soil-landscape models followed by the application of these models by GIS databases to generate predictive property maps (Petersen et al., 1991). The geographic databases of landscape information that are used to extend these models are basically Jenny's soil-forming factors where climate is expressed as slope aspect, organisms as land cover, relief as digital elevation models, parent material as a geology map and time as the geomorphic position. While we are on a learning curve in advancing our state-of-knowledge from conceptual, mental or verbal soil models of Jenny's period to the computer era of simulation models, we have just begun this challenging new frontier. There is much to be done and many opportunities along the way.

SUMMARY

The state factor model as presented by Jenny (1941) has had more impact on pedogenic studies to understand soil formation than any other soil model published in the past 50 yrs. The state factor model is pinnacle in teaching students the broad conceptual framework to comprehend generalized soil patterns. The state factor model has been widely applied to derive soil factor sequences by ordination of state factor variables. Assumptions of factor independence are generally not shared by most pedologists and may not be rigorously realized. Other major contributions of Jenny's work include: communication of Russian pioneering work, methodology for pedogenic quantification, institution of pedogenic credibility, stimulus to develop other pedogenic models, construction of Soil Taxonomy, and fostering an ecosystem approach to environmental challenges.

REFERENCES

Arnold, R.W. 1965. Multiple working hypotheses in soil genesis. Soil Sci. Soc. Am. Proc. 29:717-724.

Arnold, R.W. 1983. Concepts of soils and pedology. p. 1-21. *In* L.P. Wilding et al (ed.) Pedogenesis and soil taxonomy I. Concepts and interactions. Developments in soil science 11A. Elsevier Sci. Publ., Amsterdam.

Arnold, R.W. 1991. Soil geography and factor functionality: Interacting concepts. p. 99-109. *In* R. Amundson et al. (ed.) Factors of soil formation: A fiftieth anniversary retrospective. SSSA Spec. Publ. 33. ASA, CSSA, and SSSA, Madison, WI.

Birkeland, P.W. 1984. Soils and geomorphology. Oxford Univ. Press, London.

Bryant, R.B., and C.G. Olson. 1987. Soil genesis: Opportunities and new directions for research. p. 301-311. *In* L.L. Boersma (ed.) Future developments in soil science research. SSSA, Madison, WI.

Buol, S.W., F.D. Hole, and R.J. McCracken. 1989. SOil genesis and classification. 3rd ed. Iowa State Univ. Press, Ames, IA.

Byers, H.G., C.E. Kellogg, M.S. Anderson, and J. Thorp. 1938. Formation of soils. p. 948-978. *In* Soils and men. U.S. Gov. Print. Office, Washington, DC.

Chadwick, O.A., G.H. Brimhall, and D.M. Hendricks. 1990. From a black to gray box—a mass balance interpretation of pedogenesis. Geomorphology 3:369-390.

Chamberlain, T.C. 1897. The method of multiple working hypothesis. J. Geol. 4:837-848.

Chessworth, W. 1973a. The parent rock effect in the genesis of soil. Geoderma 10:215-225.

Chessworth, W. 1973b. The residua system of chemical weathering: A model for the chemical breakdown of silicate rocks at the surface of the earth. J. Soil Sci. 24:69-81.

Ciolkosz, E.J., W.J. Waltman, T.W. Simpson, and R.R. Dobos. 1989. Distribution and genesis of soils of the northeastern United States. Geomorphology 2:285-302.

Cline, M.G. 1961. The changing model of soil. Soil Sci. Soc. Am. Proc. 25:442-446.

Crocker, R.L. 1952. Soil genesis and pedogenic factors. Q. Rev. Biol. 27:139-168.

Crocker, R.L. 1960. The plant factor in soil formation. p. 84-90. *In* Proc. Pacific Science Congress. 9th, Bangkok, Thailand. 18 Nov. to 9 Dec. 1957. Vol. 18. Secretariat, Bangkok, Thailand.

Dijkerman, J.C. 1974. Pedology as a science: The role of data, models, and theories in the study of natural soil systems. Geoderma 11:73-93.

Dokuchaev, V.V. 1893. The Russian steppes and study of soil in Russia, it's past and present. (Translated by J.M. Crawford) Ministry of Crown Domains, St. Petersburg, Russia.

Fanning, D.S., and C.B. Fanning. 1989. Soil: Morphology, genesis and classification. John Wiley & Sons, New York.

Gerrard, A.J.W. 1981. Soils and landforms. George Allen and Unwin, Boston.

Hilgard, E.W. 1921. Soils: Their formation, properties, composition, and plant growth in the humid and arid regions. Macmillan Co., London.

Hoosbeek, M.R., and R.B. Bryant. 1992. Towards the quantitative modelling of pedogenesis—A review. Geoderma 55:183-210.

Huggett, R.J. 1975. Soil landscape systems: A model of soil genesis. Geoderma 13:1-22.

Huggett, R.J. 1976. Conceptual models in pedogenesis—A discussion. Geoderma 16:261-262.

Jenny, H. 1938. Properties of colloids. Stanford Univ. Press, Palo Alto, CA.

Jenny, H. 1941. Factors of soil formation. McGraw-Hill Book Co., New York.

Jenny, H. 1961. Derivation of state factor equations of soils and ecosystems. Soil Sci. Soc. Am. Proc. 25:385-388.

Jenny, H. 1980. The soil resource, origin and behavior. Springer-Verlag, New York.

Jenny, H. 1989. Soil Scientist, teacher and scholar. Regional Oral History Office Bancroft Library, Univ. of California, Berkeley.

Joffe, J.S. 1936. Pedology. Rutgers Univ. Press, New Brunswick, NJ.

Johnson, D.L., D. Watson-Stegner. 1987. Evolution model of pedogenesis. Soil Sci. 143:349-366.

Johnson, D.L., E.A. Keller, and T.K. Rockwell. 1990. Dynamic pedogenesis: New views of some key soil concepts and a model for interpreting quanternary soils. Quat. Res. (NY) 33:306-319.

Kellogg, C.E. 1936. Development and significance of the great soil groups of the United States. USDA Misc. Publ. no. 229. U.S. Gov. Print. Office, Washington, DC.

Kellogg, C.E. 1937. Soil survey manual. USDA, Misc. Publ. no. 274, U.S. Gov. Print. Office, Washington, DC.

Kuhn, T.S. 1970. The structure of scientific revolutions. 2nd ed. Univ. Chicago Press, Chicago.

Levine, E.R., and E.J. Ciolkosz. 1986. A computer simulation model for soil genesis applications. Soil Sci. Soc. Am. J. 50:661-667.

Levine, E.R., and E.J. Ciolkosz. 1988. Computer simulation of soil sensitivity to acid rain. Soil Sci. Soc. Am. J. 52:209-215.

Lyon, T.L., and H.O. Buckman. 1938. The nature and property of soils. 3rd ed. Macmillian, New York.

Platt, J.R. 1964. Strong inference. Science (Washington, DC) 146:347-353.

Petersen, G.W., G.A. Nielsen, and L.P. Wilding. 1992. Geographic information systems and remote sensing in land resource analysis and management. Suelo Planta 1:531-543.

Ruhe, R.V. 1969. Quaternary landscapes in Iowa. Iowa State Univ. Press. Ames, Iowa.

Ruhe, R.V. 1975. Geomorphology. Houghton Mifflin, Boston.

Runge, E.C.A. 1973. Soil development sequence and energy models. Soil Sci. 115:183-193.

Russell, E.J. 1937. Soil conditions and plant growth. 7th ed. Longmans, Green and Co., London.

Simonson, R.W. 1959. Outline of a generalized theory of soil genesis. Soil Sci. Soc. Am. Proc. 23:152-156.

Simonson, R.W. 1978. Multiple process model of soil genesis. p. 1-25. *In* W.C. Mahaney (ed.) Quat. Symp. York Univ., 3rd, Toronto, Canada. 15 to 16 May 1976. Geo Abstracts, Toronto, Canada.

Singer, M.J., and D.N. Munns. 1991. Soils: An introduction. 2nd ed. Macmillian Publ. Co., New York.

Smeck, N.E., and E.C.A. Runge. 1971. Factors influencing profile development exhibited by some hydromorphic soils in Illinois. p. 169-179. *In* E. Schlichting and U. Schwertmann (ed.) Pseudology and gleys. trans. Comm. 5, 6, Int. Soc. Soil Sci., Chemie-Verlag.

Smeck, N.E., E.C.A. Runge, and E.E. Mackintosh. 1983. Dynamics and genetic modelling of soil systems. p. 51-81. *In* L.P. Wilding et al. (ed.) Pedogenesis and soil taxonomy I. Concept and interactions. Developments in soil science 11A. Elsevier Sci. Publ. Co., Amsterdam.

Sposito, G., and R.J. Reginato (ed.). 1992. Pedology: The science of soil development. p. 9-25. *In* G. Sposito and R.J. Reginato (ed.) Opportunities in basic soil science research. SSSA, Madison, WI.

Soil Survey Staff. 1975. Soil taxonomy; a basic system of soil classification for making and interpreting soil surveys. USDA-SCS Agric. Handb. 436. U.S. Gov. Print. Office, Washington, DC.

Stephens, C.G. 1947. Functional analysis in pedogenesis. Trans. R. Soc. S. Aust. 71:168-181.

Stevens, P.R., and T.W. Walker. 1970. The chronosequence concept of soil formation. Q. Rev. Biol. 45:333-350.

Tandarich, J. 1991. The intellectual background for Factors of Soil Formtion. p. 1-13. *In* R. Amundson et al. (ed.) Factors of soil formation: A fiftieth anniversary retrospective. SSSA Spec. Publ. 33. SSSA, Madison, WI.

U.S. Department of Agriculture. 1938. Soils and men, USDA yearbook, U.S. Gov. Print. Office, Washington, DC.

Van Wambeke, A., H. Eswaran, A.H. Herbillion, and J. Comerma. 1983. Oxisols. p. 325-354. *In* L.P. Wilding (ed.) Pedogenesis and soil taxonomy II. Elsevier Scientific Publ. Co., Amsterdam.

Vepraskas, M.J., and L.P. Wilding. 1983. Deeply weathered soils in the Texas Coastal Plain. Soil Sci. Soc. Am. J. 47:293-300.

Vitousek, P. 1991. Factor controlling ecosystem structure and function. p. 87-97. *In* R. Amundson et al. (ed.) Factors of soil formation: A fiftieth retrospective. SSSA Spec. Publ. 33. SSSA, Madison, WI.

Wilding, L.P. 1986. Highlights of activities in division S-5—soil genesis, morphology, and classification. Soil Sci. Soc. Am. J. 50:1377-78.

Wilding, L.P. 1988. Improving our understanding of the composition of the soil-landscape. p. 259. *In* Proc. of an Int. Interactive Workshop on Soil Resources: Their inventory, analysis and interpretation for use in the 1990's. Educational Dev. System, Minnesota Ext. Service, Univ. of Minnesota, St. Paul.

Wilding, L.P., and K. Flach. 1985. Micropedology and soil taxonomy. p. 1-16. *In* J.M. Bigham et al. (ed.) Soil micromorphology and soil classification. SSSA Spec. Publ. 15. SSSA, Madison, WI.

Wilding, L.P., and L.R. Drees. 1983. Spatial variability and pedology. p. 83-116. *In* L.P. Wilding et al. (ed.) Pedogenesis and soil taxonomy: I. Elsevier Publ. Co., Amsterdam.

Wilding, L.P., N.E. Smeck, and G.F. Hall. (ed.). 1983a. Pedogenesis and soil taxonomy I. Elsevier Sci. Publ. Co., Amsterdam.

Wilding, L.P., N.E. Smeck, and G.F. Hall. (ed.). 1983b. Pedogenesis and soil taxonomy II. Elsevier Sci. Publ. Co., Amsterdam.

Yaalon, D.H. 1971. Paleopedology—origin, nature and dating of paleosols. Int. Soc. Soil Sci. and Israel Univ. Press, Jerusalem.

Yaalon, D.H. 1975. Conceptual models in pedogenesis. Can soil-forming functions be solved? Geoderma 14:189-205.

3 The Environmental Factor Approach to the Interpretation of Paleosols

Gregory J. Retallack

University of Oregon
Eugene, Oregon

ABSTRACT

Paleosols ranging in geological age to 3 billion years are widely used as evidence for ancient surface environments—an enterprise dependent on the enormous literature on factors in soil formation, popularized by Hans Jenny. This kind of inference inverts the logic of Jenny in a manner common to geological sciences—deducing paleoenvironments from observed paleosol features, rather than deducing variation in soil features with observed environmental differences. A paleosol is a single product of many past influences, including alteration after burial, and because of this, some environmental relationships with soil color, clayeyness and organic matter are not useful for interpreting paleosols. One relationship that has proven useful for paleosols is that between depth to calcic horizon and mean annual rainfall. A new compilation of data presented here demonstrates that this relationship holds for aridland soils worldwide. The use of this relationship for interpreting paleoclimate from paleosols is illustrated with an example of the Eocene and Oligocene paleosols of Badlands National Park, South Dakota. Other approaches for the study of paleosols include identifying paleosols within a soil taxonomy, and simulating ancient soil development with mathematical process models. Identification of paleosols leads to broad areas on soil maps unless done with several paleosols. Process models often founder on assumptions, and those of the form $\delta x/\delta t$ (where x is a measured soil property) are difficult to apply because time of formation (t) of a paleosol is estimable only to an order of magnitude. Thus the environmental factor approach to the interpretation of paleosols is likely to remain popular for some time to come.

Hans Jenny's book *Factors in soil formation* (1941) was seminal in systematizing and quantifying empirical studies of soil formation. One can argue that his approach was not fundamentally distinct from that implied by Dokuchaev (1883) and other pioneers of soil science, or that the full promise of Jenny's universal equation has yet to be realized (Yaalon, 1975). Nevertheless, Jenny's vision of quantitative study of soil formation by means of soils chosen

---

Copyright © 1994 Soil Science Society of America, 677 S. Segoe Rd., Madison, WI 53711, USA. *Factors of Soil Formation: A Fiftieth Anniversary Retrospective*. SSSA Special Publication 33.

according to an explicit experimental design continues to have an undeniable influence in soil science, as well as in geological studies of soils and paleosols (Birkeland, 1984). Geological sciences include theoretical and taxonomic studies like those of soil science, but Jenny's approach is most similar to other kinds of study, such as the functional morphology of mammals. In such studies are documented the influence of specific factors such as diet on tooth and limb structure of mammals (Dodd & Stanton, 1990). Data to document such relationships needs to be chosen with care to mitigate other possible causes of variation. In Jenny's terms such a carefully chosen data set is analogous to a biosequence, and its study would aim at establishing biofunctions.

This however is only part of the task for geological sciences, which address not only how the world works at present, but what such understanding can reveal about events in the past. For example, with the knowledge that modern grassland mammals have high crowned teeth and elongate slender limbs with few digits, grasslands of the past can be inferred from finding such teeth and limb bones. Uniformitarianism is the term applied in geological sciences for such inferences about the past based on observations of the present. There is nothing special to geology about such logic. Indeed, it is little more than a kind of natural controlled experiment (Shea, 1982), or making inferences about the unknown (in this case events of the past) based on the known (observations of the present). It is this kind of logic that makes the environmental factor approach to the study of soils so useful for the interpretation of paleosols buried within sedimentary rock sequences ranging back in geological age up to 3 billion years (Retallack, 1990). Ongoing research on the relationships between soil features and the natural environment provides a large fund of potential inferences about the past from features observed in paleosols (Birkeland, 1984). This uniformitarian use of these data does not in itself attempt to extend or improve factor functions, although their limitations must be considered, but rather to use them. Such uniformitarian interpretation of paleosols also has limitations. Many details of soil formation cannot be observed in paleosols altered by burial over the millions of years since they formed (Retallack, 1991a).

The establishment of relationships between environment and features of surface soils and the use of such relationships to interpret paleosols are two distinctly different kinds of study in aims, materials, methods, scale and resolution. In this essay they will be illustrated by a specific example of the relationship between mean annual rainfall and depth to calcareous nodules in soils and paleosols. This is just one of numerous soil–environment relationships useful for interpreting paleosols, and only a tabular summary of other relationships can be offered here. None of these are without problems for application to paleosols, but then neither are taxonomic and modeling approaches, which also will be discussed as alternatives to the environmental factor approach.

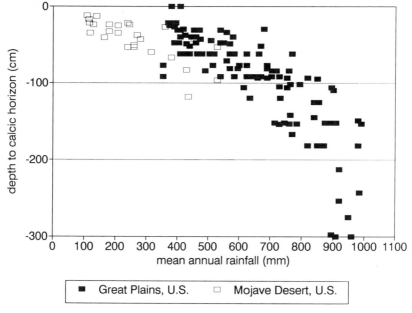

Fig. 3-1. The relationship between depth to calcic horizon in soils and their mean annual precipitation for the North American Great Plains (Jenny & Leoanrd, 1935; Jenny, 1941) and Mojave Desert (Arkley, 1963).

DEPTH TO CALCIC HORIZON AND MEAN ANNUAL RAINFALL

Previous Work

One of the most impressive quantifications of factors in soil formation in Jenny's (1941) book and an earlier paper (Jenny & Leonard, 1935) was a scatter plot (Fig. 3-1) of the depth to the zone of calcareous nodules in soils vs. mean annual rainfall, with the linear relationship $D = 86.74 - 0.2835P$, where D is the depth (cm) down from the surface to the top of the calcic horizon of a soil, and P is the mean annual precipitation (mm) at that site. I have replotted these data in metric units from Jenny's plot of 106 soils, giving a correlation coefficient (r) of 0.81 and a standard error (s) about the regression of ± 22 cm, using computer routines of Davis (1973). This relationship is an excellent example of a climofunction in the sense of Jenny (1941) because not only other aspects of climate, but also vegetation, topographic relief, parent material and time were more or less constrained for the set of soils considered. The relevant soils were on an east-west transect across postglacial (less than 15 000 yr old) loess of rolling grasslands of the central North American Great Plains.

Several subsequent studies have refined this work. Ruhe (1984) examined another transect of soils in this same area, and faulted Jenny's relationship because the time for formation and vegetation history at either ends of the transect were not exactly comparable. The eastern Udolls in the transect formed on Peoria Loess some 14 000 yr old, whereas the western Ustolls formed on Bignell Loess no more than 9 000 yr old. The eastern soils also were forested for a part of their postglacial history, unlike the western soils. Arkley (1963) also has been critical of this relationship because in the Mojave Desert he found a different relationship of $D = -2.734 - 0.1508P$. I also have replotted this data set of 26 soils, and calculated a correlation coefficient (r) of 0.75 and standard error (s) of ±17 cm. The difference between the relationship of Jenny and Arkley is not so profound as apparent from casual inspection of the equations. Mojave and Great Plains data overlap substantially (Fig. 3-1), and may be better fit by a curve as shown schematically for Israeli soils by Dan and Yaalon (1982) and Yaalon (1983). The difference may be due to winter rainfall in the southwestern USA rather than summer rain of the Midwest, and to differences in water movement through the soil related to differences in porosity and structure of the soils in the two areas (Birkeland, 1984). Such differences also explain the differing values of precipitation across the udic–ustic soil moisture boundary in different countries (Yaalon, 1983).

Another group of soils from the southwestern USA was used by Marion et al. (1985) to calibrate a computer model of the depth of carbonate in aridland soils. They concluded that carbonate depths reflected a slightly more humid climate of the past rather than present mean annual rainfall. Subsequent computer modeling studies have shown that the depth of carbonate in soils is controlled by a complex mix of soil porosity, rainfall infiltration rates, soil respiration, available Ca, dust influx, temperature and other factors (McFadden & Tinsley, 1985; Mayer et al., 1988; McFadden et al., 1991). Because biological productivity also is related to mean annual rainfall and to soil respiration rates (Leith, 1975), the depth to the calcic horizon also can be used as a proxy indicator of rangeland productivity (Munn et al., 1978).

A New Compilation

From the perspective of high-resolution Quaternary studies these climofunctions need to be used cautiously. From the perspective of low-resolution geological interpretation of ancient paleosols; however, it is impressive how such a relationship could emerge at all from such a complex natural phenomenon. Indeed there have been indications in the literature that this relationship between depth to carbonate nodules and rainfall holds in such different soil-forming environments as the Serengeti Plains of Tanzania (de Wit, 1978), the Indo-Gangetic Plains of India (Sehgal et al., 1968) and the deserts of Sinai and Negev (Dan & Yaalon, 1982). These hopeful signs stimulated me to compile as many data as could be found in the literature on depth to the calcic horizon of soils whose mean annual precipitation also is known. The data set now stands at 317 soils, and includes soils from the Great Plains of North America (Jenny & Leonard, 1935; Jenny, 1941; Ruhe, 1984); the desert Southwest of the USA (Arkley, 1963; Marion et al., 1985); the pampas of Argentina (Fadda, 1968; U.N.FAO, 1971; Soil Correlation Committee for South America, 1967; Plaza & Moscatelli, 1989); the central Spanish meseta (del Villar, 1957); the Serengeti Plains of Tanzania (de Wit, 1978; Jager, 1982); the Kalahari Desert of Botswana (Siderius, 1973); the Sinai-Negev deserts of Israel (Dan et al., 1981; Dan & Yaalon, 1982); the Mesopotamian Plains of Syria, Iraq, and Iran (Mulders, 1969; Al Taie et al., 1969; U.N.FAO, 1977; Hussain et al., 1984); the Indo-Gangetic Plains of India (Sehgal et al., 1968; U.N.FAO, 1977; Sidhu et al., 1977; Ahmad et al., 1977; Bhargava et al., 1981, Vinayak et al., 1981, Murthy et al., 1982; Courty & Féderoff, 1985); the south Russian Plain (Dokuchaev, 1883; Glinka, 1931; U.N.FAO, 1981); the southwestern Siberian Plain (Fedorin, 1960; Bal & Buursink, 1976); the western Mongolian Plain (Nogina, 1976); the north China Plain (Thorp, 1936; Bronger & Heinkele, 1989); the Riverina district of central southeastern Australia (Stace et al., 1968); intermontane basins of the South Island of New Zealand (McCraw, 1964; Raeside & Cutler, 1966; Leamy & Sanders, 1967; Soil Bureau Staff, 1968; Orbell, 1974); and the polar deserts of Antarctica (Campbell & Claridge, 1987) and Greenland (Tedrow, 1970, 1977). Where climatic data were not provided within these soil studies they were obtained from published tabulations and isohyet maps (Alt, 1932; Watts, 1969; Hoffman, 1975; Lydolph, 1977; Taha et al., 1981; Ruffner, 1985).

A few ground rules were used for selection of soils for this database, which is presented here in full (Appendix 3-1) in the hope that others will expand upon it. All are late Pleistocene and Holocene soils on unconsolidated sediments other than clay or limestone in low lying or rolling terrain of free drainage, more or less as in Jenny and Leonard's (1935) original data. The calcic horizon was taken as the horizon of most abundant micritic carbonate nodules that appear to be in place. This was not always the highest nodule or carbonate in a profile, especially within monsoonal soils which may have at least a few calcareous nodules throughout the profile (Retallack, 1991b). Although soil horizons above the calcic horizon may be weakly calcareous because leached of carbonate, this measure of the depth to the horizon of carbonate acccumulation is not the same as the depth of leaching of a calcareous parent material, which apparently reflects time for formation rather than climatic conditions (Birkeland, 1984). Carbonate "veins" or "cement" were not accepted as a calcic horizon, because these could be of groundwater origin, rather than pedogenic origin (Mann & Horwitz, 1979; Carlisle, 1983; Lander, 1990; Kaemmerer & Revel 1991; Wright & Tucker, 1991). Solid carbonate layers ("K horizons" of the USA or "tosca" of Argentina) also were not included because they indicate soils of great antiquity and more complex history (Gile et al., 1966; Pazos, 190), nor were soils with stone lines or other indications of redeposition in alluvial parent materials (Courty & Féderoff, 1985). Also not included were soils of hills and steep slopes, nor soils on bedrock, limestone, beach rock, obvious local sand dunes or clay. In most cases, these various complications were apparent from micromorphological, geochemical or other studies of the soils, but there are bound to be suspect soils that evaded detection. A literature compilation such as this is no substitute for a careful selection and study of all the soils by a single investigator, as can be seen from the more highly correlated results of Jenny and Leonard (1935) and Arkley (1963).

Despite these strictures, there remains considerable variation in environmental factors within the data of my compilation. Climate, for example, ranges from frigid to tropical, and from mildly seasonal to monsoonal. Most of the soils support grassland, but many are under dry woodland and desert shrubland. Their topographic setting ranges from extensive alluvial plains to loess-mantled rolling terrain. Parent materials include volcanic ash, quartzofeldspathic silty alluvium and boulder till. Time for formation varies from late Pleistocene (perhaps 50 000 yr old) to late Holocene (perhaps as young as 500 yr).

Even with those unwieldy data the points define a clear relationship between depth to calcic horizon and mean annual rainfall (Fig. 3-2). The relationship is not linear, but best-fit by the curve $D = -40.49 - 0.0852P - 0.0002455P^2$, which has a correlation coefficient (r) of 0.78 and a standard error (s) about the curve of ± 33 cm. Application of an F test to an analysis of variance (following methods of Davis, 1973) shows that this fit is highly significant at the 1% level, and is at least an equally significant advance over the fit provided by linear regression. The fitting of higher-order polynomial curves however, gave insignificantly improved fit. This is a surprisingly clear relationship for a data set of such diverse origins and it encourages faith that something fundamental about the way soils form has emerged above the noise of local variations in soil history and other factors. Better results have been obtained locally by Jenny and Arkley with more carefully and deliberately constrained field studies. Such studies remain a promising direction for further research, and can be constructed from the raw data of Appendix 3-1.

Problems in Application to Paleosols

From the perspective of interpreting paleosols, not the relationship of depth of carbonate to rainfall, but rather the relationship of rainfall to depth of carbonate is needed. Regression of this relationship for the new compliation yields the equation $P = 139.6 - 6.388D - 0.01303D^2$, with a correlation coefficient (r) of 0.79 and a standard error (s) of ± 141 mm. The broad error envelope is in part a consequence of compression of the data toward the soil surface (Fig. 3-2). This level of resolution may not be helpful for studies of late Quaternary paleosols, but can be valuable for paleoclimatic interpretation of paleosols in ancient sedimentary rocks. Furthermore, there is the prospect of interpreting paleoclimate from subsets of these data—climosequences constrained more narrowly to approximate other paleoenvironmental conditions of particular paleosols. For example, in the Great Plains data of Jenny and Leonard (1935) and Jenny (1941) the relationship is $P = 418.5 - 2.335D$, with a correlation coefficient of 0.81 and a standard error of ± 108 mm.

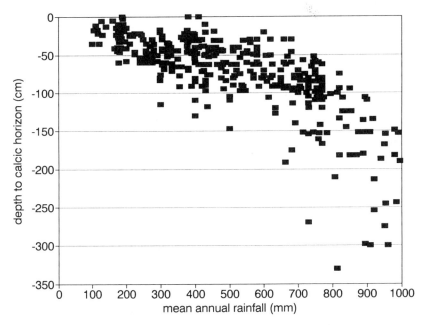

Fig. 3-2. The relationship between depth to calcic horizon in soils and their mean annual precipitation from a new compilation of 317 aridland soils from all over the world (see text for sources).

There are troublesome, but not unsuperable, additional problems in applying these data to the interpretation of paleoclimate from paleosols, because of different atmospheric concentrations of CO_2 during the geological past, the erosion of soils before burial and their compaction after burial. The first of these is difficult to circumvent because there are no soils presently forming under the kind of global greenhouse climate of elevated atmospheric CO_2 envisaged, for example, during mid-Cretaceous geological time. Indeed, Ice Ages like that which have persisted for roughly the past 2 million years are comparatively rare events in earth history. In the distant geological past these Ice Ages with their waxing and waning of large continental ice sheets have persisted for durations only a few tens of millions of years, some 290 million years ago (Carboniferous–Permian boundary), again some 440 million years ago (Ordovician–Silurian boundary), 600, 1000, and 2300 million years ago (all Precambrian; Hambrey & Harland, 1981). Analysis of gas bubbles in ice cores indicates that CO_2, which is currently at levels of 300 mg L^{-1}, may have been as high as 400 mg L^{-1} during a warm climatic period a few thousands of years ago, and as low as 200 mg L^{-1} during the glacial maximum, 15 000 yr ago (Neftel et al., 1982). Isotopic study of paleosol carbonates combined with a theoretical model for the C isotopic systematics of soils have been used as evidence that atmospheric CO_2 was no more than 700 mg L^{-1} during the Eocene, but perhaps on the order of 1500 to 3000 mg L^{-1} during Early Cretaceous time (Cerling, 1991). This order of magnitude variation in atmospheric CO_2 abundance is less than the two orders of magnitude difference between CO_2 abundance in the atmosphere and in some modern highly productive soils (Brook et al., 1983), but during such greenhouse times in earth history rainwater would have been somewhat more acidic. Thus it is possible that the level within soils at which nodules formed may have been deeper for any given mean annual precipitation. How much deeper for a given level of CO_2 is difficult to calculate, because the calculations are dependent on estimates of soil respiration. Estimates can be made using the numerical models of McFadden and Tinsley (1985) and Mayer et al. (1988).

The problems of erosion of paleosols before burial is less troublesome. Many soils eroded due to misguided agricultural practices of the past are included in the new compilation (Fig. 3-2). In addition, surface horizon structures in many aridland paleosols are distinctive in the field from those in their subsoil horizons. The dark color and organic matter of mollic epipedons are seldom preserved in buried soils that were well drained, but there are commonly granular and platy peds defined by argillans and a variety of root traces and rhizoconcretions (Retallack, 1990, 1991a,b). These can be examined in the field, along with the nature of the contact between the paleosol and its overlying sediments, to determine the likelihood that depth to carbonate nodules has been reduced by erosion of the soil. There is not much that can be done to reconstruct severely eroded paleosols, and they are best ignored for the purpose of paleoclimatic reconstruction. In general, surface horizons of paleosols are widely and well preserved in paleosols within thick sedimentary sequences that accumulated in subsiding sedimentary basins, but compound paleosols with eroded topsoils are widespread in tectonically stable or uplifted continental regions (Retallack, 1986a; Schaetzl & Sorenson, 1987).

A final difficulty is compaction of paleosols due to loading of overburden after burial and to other kinds of tectonic deformation. The degree of compaction of paleosols as geologically young as Miocene can be significant if, for example, they were buried by large volcanic edifices or in such thick sedimentary sequences as Himalayan outwash (Retallack, 1991b). The best direct evidence for degree of compaction is what in geological sciences is called a clastic dike, and in soil sciences is called a silan; that is to say, a former near-vertical crack in the paleosol that has become filled with a contrasting material that buried the paleosol. Such structures become tightly folded with compaction of the surrounding paleosol matrix, and have been useful in reconstructing the original thickness of paleosols as geologically ancient as Precambrian (Retallack, 1986b). Root traces and burrows in paleosols also become folded and flattened with compaction of paleosols, but in these cases original irregularity and flattening diminishes their usefulness as indicators of compaction during burial. Another approach is to determine how much higher is the bulk density of paleosol samples than of samples of comparable surface soils (Retallack, 1991b). Density of paleosols and soils is quite variable and difficult to measure with sufficient accuracy for this to be more than a check on the results of other approaches. The compaction correction method that I prefer is to reconstruct depth of burial from geological data and then estimate likely compaction from standard curves for the compaction of appropriate kinds of rocks with depth (Baldwin & Butler, 1985). Reconstructing the depth of burial may be possible by considering regional stratigraphy and structure, but there also are a variety of other indices of burial including vitrinite reflectance, hydrocarbon maturation, and pollen color alteration (Tissot & Welte, 1984).

Example of Eocene-Oligocene Paleosols and Paleoclimate

A specific case study of the paleoclimatic interpretation of Eocene and Oligocene paleosols in Badlands National Park, South Dakota, may clarify what the factor-function approach can tell us about the past. The Badlands are well known as a long ragged wall of colorful clayey nonmarine rocks between high prairie to the north and the valley of the White River to the south (Fig. 3-3). They have long been famous as the foremost example of badlands weathering (Schumm, 1975) and for a great variety of well-preserved fossil mammals (Emry et al., 1987) now known to range in age from late Eocene to early Oligocene (somewhat older than previously thought; Swisher & Prothero, 1990). My own studies of the numerous paleosols within this sedimentry succession have focused on a single measured section in the Pinnacles area of the National Park (Fig. 3-4), which has been used as evidence of grassland ecosystems (Retallack, 1982, 1984b, 1988a, 1990), the completeness of stratigraphic sections (Retallack, 1984a), factors in sedimentation (Retallack, 1986a), and paleoclimatic change (Retallack, 1992a).

Many of these paleosols are studded with calcareous nodules that have all the earmarks of being original soil nodules. They form definite horizons below the surface horizons of the paleosols and some show mild surface ferruginization. In petrographic thin section, micritic matrix can be seen to locally replace pre-existing grains and local sparry calcite displaces soil peds (soil clods) and fills hollows after root traces (Retallack, 1983a). Similar nodules are commonly found as clasts in channel deposits at the same stratigraphic level (Wanless, 1923). Fossil skulls, tortoises (*Stylemys nebrascensis*), and other hollow groups of bones are noticeably less crushed, distorted and oxidized within nodules than in adjacent clayey paleosol matrix, and so predate lithostatic compaction. Nodules also contain more weatherable minerals, such as volcanic glass, hypersthene and olivine, than the surrounding claystone. Finally, some nodules are penetrated by drab-haloed root traces, thought to have formed during the early burial decomposition of the last crop of roots before burial of the paleosols (Retallack, 1983a, 1991a).

Although there are some clearly pedogenic nodules in Oligocene paleosols of these sequence, there also are carbonate layers and a widely dispersed carbonate cement. This latter cement is most conspicuous within paleochannel sandstones and in C horizons of paleosols in the silty upper part of the sequence, and is evidence of a component of carbonate added to the sequence presumably from groundwater during burial. The carbonate layers found mainly within the late Eocene clayey paleosols are more problematic. They formed within the soil zone, because they include replacive micritic matrix,

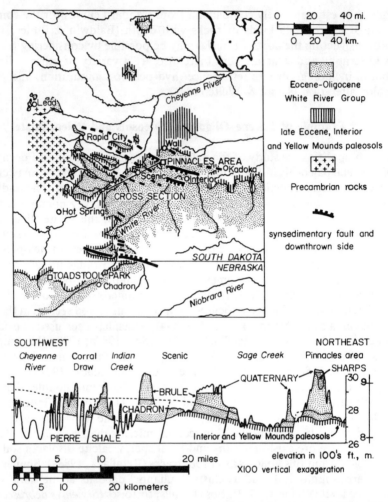

Fig. 3-3. Location (above) and geological cross section (below) of a measured section of Eocene and Oligocene paleosols in the Pinnacles area of Badlands National Park, South Dakota (from Retallack, 1988a).

displacive brecciated textures, cellularly preserved root traces, calcite crystal tubes after roots, fossil larval cells of dung beetles (*Pallichnus dakotensis*) and bees (*Celliforma ficoides*), and displacive sparry calcite veins. They also are disrupted by fossil root traces (Retallack, 1983a, 1984b). However, their morphology is unusual for soils. There may be up to five of these calcareous stringers within a single profile, each stringer with sharp upper and lower contact. Although roughly stratabound, the stringers curve up and down across bedding and thicken and thin irregularly. I have long been uncom-

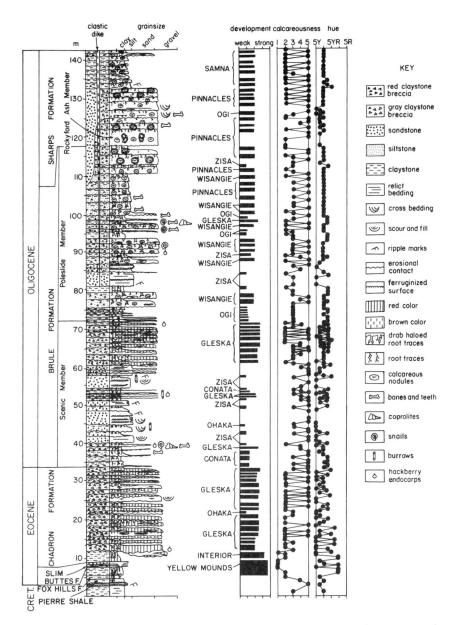

Fig. 3-4. A measured section of Eocene and Oligocene paleosols from the Pinnacles area of Badlands National Park, South Dakota (from Retallack, 1990). Individual paleosols are marked by heavy bars, the width of which corresponds to relative degree of soil development. Calcareousness is from a nominal scale of Retallack (1988b) and hue from a Munsell color chart compared with freshly excavated rock.

fortable interpreting these as strongly developed calcic (K) horizons (Retallack, 1983a), and new isotopic results obtained by Lander (1990) have convinced me that these are near-surface examples of valley calcretes, similar to those now found in the deserts of Western Australia (Mann & Horwitz, 1979) and Namibia (Carlisle, 1983). Also, like these valley calcretes the Badlands petrocalcic stringers are most prolific near paleochannel sandstones and include patches of opal, chalcedony and carnotite. There is little theoretical reason to suspect that carbonates derived in part from groundwater evaporation within the soil zone should show the same relationship to mean annual rainfall as carbonate nodules of well-drained aridland soils; however, the two modern cases are very near the surface of the soil in very dry regions. The paleoclimatic significance of valley calcretes remains to be established.

Paleoclimatic interpretation of calcareous paleosols of Badlands National Park is not severely compromised by changing atmospheric levels of CO_2. The Eocene-Oligocene boundary was a time of global climatic cooling, the most significant of several abrupt climatic deteriorations which marked the transition from a middle Eocene greenhouse climate to the Plio-Pleistocene Ice Age (Miller, 1992; Retallack, 1992a). Calcareous paleosols of Badlands National Park are principally of Oligocene age and probably formed under atmospheric CO_2 levels much less than 700 mg L^{-1} postulated for the Eocene by Cerling (1991) and closer to interglacial levels of about 300 to 400 mg L^{-1} experienced by many surface soils of the new compilation (Fig. 3-2).

Erosion of the surface of paleosols in the Badlands before their burial also was minimal. Few of the paleosols in this sequence of wind and water redeposited volcanic ashes (Evanoff et al., 1992) are sharply overlain by sediments, and those that are sharply truncated also have preserved the platy to granular structure and fine root traces of their surface horizons (Retallack, 1983a, 1990).

Compaction of the paleosols also was not a problem for interpretation of the depth of their carbonate nodules. The maximum thickness of the Eocene-Oligocene White River Group is 195 m, and the cumulative thicknesses of other Oligocene and geologically younger formations in this region total 361 m (Martin, 1983). If a marine, noncalcareous shale was buried to a depth of 500 m, it would be compacted to about 90% of its former thickness (Baldwin & Butler, 1985). This seems a very unlikely maximal compaction even at the base of the sequence, because the geologically younger sediments did not form a single layer, but filled local paleovalleys excavated as the nearby Black Hills and Great Plains continued to rise (Schultz & Stout, 1980; Swinehart et al., 1985; Angevine & Flanagan, 1987). This together with the increasingly pervasive calcareous cement and silty texture higher in the sequence makes it unlikely that compaction was significant.

The depths to calcareous nodular horizons within paleosols, recorded in the field during section measuring, show a decline in paleosols stratigraphically higher in the section that, from the relationship demonstrated for surface soils (Fig. 3-2), can be interpreted as evidence of Oligocene climatic drying (Fig. 3-5). The petrocalcic horizons, for which paleoclimatic sig-

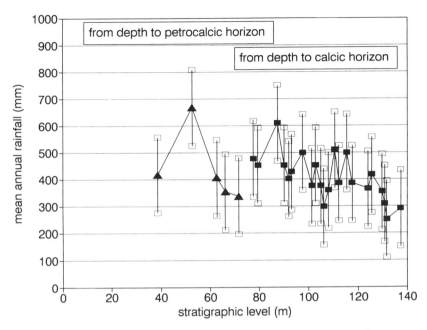

Fig. 3-5. Higher, and thus geologically younger, within the measured section of Eocene and Oligocene paleosols from South Dakota (Fig. 3-2), calcic horizons become shallower within the profile. Using the relationship compiled here (Fig. 3-1), this can be interpreted as a result of declining mean annual rainfall during Oligocene time. Closed triangles and boxes are calculated rainfall using regression and open boxes are error ($\pm 1\sigma$).

nificance has yet to be demonstrated convincingly, also continue the trend of wetter climate further back into the Eocene where the paleosols are largely noncalcareous. The basal paleosol of the sequence is leached of carbonate in its calcareous marine sedimentary parent material to a depth of 5 m. This deeply leached Eocene paleosol may have formed in a climate with annual rainfall well in excess of 1000 mm. Rainfall enjoyed by overlying late Eocene paleosols is uncertain, but was probably intermediate between this and early Oligocene paleosols for which mean annual rainfall was some 450 mm ± 141. Additional drying during mid-Oligocene time brought rainfall down to semiarid levels of about 350 mm ± 141. There is great fluctuation in the individual data points, much of which is not significant given the large standard error from the compilation of modern soils. However, the especially marked breaks at 72 and 118 m coincide with both local periods of gully erosion and nondeposition (Retallack, 1986a) and global episodes of climatic deterioration (Retallack, 1992). Within each depositional unit punctuated by erosional and climatic events the trend revealed by linear regression through the fluctuation is in each case a climatic drying (Retallack, 1986a, 1992).

Such evidence from paleosols can be important additions to other paleoclimatic indicators, such as eolian sediments, fossil leaves, and vertebrates. Drier climate of later Oligocene time also may explain the decreased abundance of kaolinite and increased abundance of smectite and then illite in paleosols higher in the sequence, as well as changes in hue from red to brown and then yellow, and changes in texture from clayey to silty (Retallack, 1986a). Climatic drying and possibly also cooling is prominently displayed in the appearance of these colorfully banded badlands. Although few fossil plants are preserved in the Badlands of South Dakota, the paleosols contain abundant drab-haloed root traces, and their profile form and soil structures also are evidence of changing vegetation; moist forests of the Eocene [37 millions of years before the present (M.Y.B.P.)] giving way to dry forests by late Eocene (35 M.Y.B.P.), to dry woodland by Oligocene (33 M.Y.B.P.), to wooded grasslands with streamside gallery woodland by 32 M.Y.B.P. and large areas of open grassland by 30 M.Y.B.P. (Retallack, 1983a, 1986a). Some changes in assemblages of fossil vertebrates of Badlands National Park can be related to progressively drier conditions and more open vegetation. Climatic drying at the Eocene–Oligocene boundary correlates in time with local extinction of large titanotheres, alligators and a wide variety of turtles and amphibians (Hutchison, 1982; Emry et al., 1987). The long-term evolutionary trend toward the kinds of cursorial limbs and high-crowned teeth well displayed in modern mammalian faunas of grasslands was initiated during this time, but in these respects Eocene and Oligocene faunas remained more like modern faunas of woodland than those of grassland (Retallack, 1990). Mid-Tertiary paleoclimatic cooling and drying was an early impetus for evolution of the grassland biome in continental interiors (Retallack, 1992a).

OTHER FACTOR-FUNCTIONS USEFUL FOR PALEOPEDOLOGY

Numerous other relationships between soil features and environmental factors also are useful for intrepreting paleoenvironments from paleosols. Only a brief tabular summary (Tables 3-1, 3-2, 3-3, 3-4, 3-5) can be offered here, but a more extended textbook treatment of the application to paleosols of these other relationships is available elsewhere (Retallack, 1990). Documentation of these relationships in surface soils can be found in a vast literature, as well as in widely available textbooks (Buol et al., 1989; Birkeland, 1984). The references to studies of surface soils cited here (Tables 3-1 to 3-5) have been chosen with deliberate preference for studies using micromorphological and bulk geochemical approaches, because these approaches have proven especially useful in the study of paleosols (Retallack, 1983a, 1991b).

Several widely discussed relationships for surface soils have been omitted from this tabular summary for use with paleosols, because their application is severely compromised by alteration of paleosols after burial, or because the measured soil features are controlled by too many factors to be teased apart. Soils buried in sedimentary successions suffer many of the same alterations as have been well documented for their enclosing sediments (Scholle & Schluger, 1979; Tissot & Welte, 1984; Surdam & Crossey, 1987). Many

ENVIRONMENTAL FACTOR APPROACH & PALEOSOLS   45

Table 3-1. Paleoclimatic indicators in paleosols.

Climatic variable	Paleosol feature	Effective range	Studies of surface soils	Application to paleosols
Mean annual temperature	Tower karst	More than 12°C	Jennings, 1985	Leary, 1981
	Cavernous subsoil weathering	More than 12°C	Jennings, 1985	Wright, 1981
	Black phytokarst	More than 12°C	Folk et al., 1973	Folk & McBride, 1976
	Spherical micropeds	More than 8°C	Stoops, 1983	Retallack, 1991b
	Pingos	Less than −1°C	Washburn, 1980	Williams, 1986
	Ice-wedge polygons	Less than −4°C	Washburn, 1980	Williams, 1986
	Sand-wedge polygons	Less than −12°C	Washburn, 1980	Williams, 1986
Mean annual precipitation	Presence of gibbsite	More than 800 mm	Sherman, 1952	Keller et al., 1954
	Proportion of smectite to kaolinitic clay	±250 mm over range 100–3000 mm	Barshad, 1966	Retallack, 1983a
	Presence of carbonate	Less than 1000 mm	Birkeland, 1984	Retallack, 1983a
	Depth to calcareous nodular (Bk) horizon	±141 mm over range 100–1000 mm	Jenny, 1941; Arkley, 1963; Retallack, 1994 (this paper)	Retallack, 1983a, 1991b
	Presence of palygorskite	Less than 400 mm	Singer & Galan, 1984	Watts, 1976
	Presence of gypsum	Less than 300 mm	Birkeland, 1984	West, 1975
	Depth to gypsum crystal (By) horizon	±100 mm over range 0–300 mm	Dan & Yaalon, 1982	West, 1975
Seasonality of rainfall	Intergrown ferric concretions and calcareous nodules	Monsoonal wet-dry seasonality	Sehgal et al., 1968; Courty & Federoff, 1985	Retallack, 1991b
	Dikes and festooned shear planes (mukkara)	Pronounced dry season, semiarid to subhumid	Paton, 1974	Allen, 1986a; Retallack, 1986b
	Common charcoal	Pronounced dry season, semiarid to humid	de Castri et al., 1981	Harris, 1957; Retallack & Dilcher, 1981
	Surface root mat and very deep sinker roots	Pronounced dry season, arid to subhumid	van Donselaar-ten Bokkel Huinink, 1966; Rutherford, 1982	Retallack, 1983a, 1991b

Table 3-2. Paleosol features of different vegetation types.

Plant formation	Root traces	Soil horizon sequence	Other soil features	Soil type	Studies of surface soils	Application to paleosols
Rain forest	Large and small in tabular mat	A–(E)–Bt–C	Spherical micropeds, few weatherable minerals, bases/alumina ratio near zero	Oxisol, Ultisol	Sanchez & Buol, 1974	Retallack, 1991c
Oligotrophic forest	Large and small, deeply penetrating	A–(E)–Bs–C	Quartz-rich, few weatherable minerals	Spodosol, Dystrochrept	Mokma & Vance, 1989	Retallack, 1990
Forest and woodland	Large and small, deeply penetrating	A–(E)–Bt–C	Blocky peds, clay skins, some weatherable minerals	Alfisol, Ultisol	Ciolkosz et al., 1990	Retallack, 1983a, 1991b
Dry woodland	Large and small deeply penetrating	A–Bt–Bk–C	Blocky peds, clay skins, common weatherable minerals and carbonate	Alfisol, Mollisol, Aridisol	Murthy et al., 1982	Retallack, 1983a, 1991b
Wooded grassland	Abundant fine (<2 mm), few large, deeply penetrating	A–(Bt)–Bk–C	Granular peds at surface, shallow calcic horizon	Mollisol, Inceptisol	de Wit, 1978	Retallack, 1991b
Open grassland	Abundant fine (<2 mm) near surface	A–(Bt)–Bk–(By)–C	Granular peds at surface, shallow calcic horizon	Mollisol, Inceptisol	Aandahl, 1982; Bronger & Heinkele, 1989	Retallack, 1983a, 1991b
Desert scrub	Sparse large, deeply penetrating	A–(Bt)–Bk–(By)–C	Vesicular, platy or blocky structure at surface, very shallow calcic horizon	Aridisol, Inceptisol, Entisol	Dan & Yaalon, 1982	Loope, 1988
Desert shrubland	Sparse, medium, woody, deeply penetrating	A–(Bt)–Bk–(By)–C	Vesicular, platy or blocky surface, very shallow calcic horizon, stone pavement	Aridisol, Inceptisol, Entisol	Stace et al., 1968	Loope, 1988

(continued on next page)

ENVIRONMENTAL FACTOR APPROACH & PALEOSOLS

Microbial earth	None		A–C	Ministromatolites, laminar crusts, claystone breccias	Entisol	Friedmann et al., 1967	Retallack, 1986b, 1990
Microbial rockland	None		A–C	Rock varnish, endolithic microrelief	Entisol	Folk et al., 1973; Friedmann & Weed, 1987; Viles, 1987	Beeunas & Knauth, 1985
Brakeland	Sparse rhizomes and fine roots		A–(Bk)–C	Platy and blocky peds at surface	Entisol, Inceptisol	Walker & Peters, 1977	Retallack, 1990
Polsterland	None		A–(Bk)–C	Platy peds, surface erosion mounds and swales	Entisol, Inceptisol	Brown & Veum, 1974	Feakes & Retallack, 1988; Retallack, 1990
Swamp	Large, in tabular mat		O–A–(Bg)–C	Woody peat or coal, noncalcareous	Histosol, Aqualf, Aquult	Ho & Coleman, 1967; Lytle, 1968	Retallack & Dilcher, 1981
Marsh	Fine, in tabuler mat		O–A–(Bg)–C	Herbaceous peat or coal, noncalcareous	Histosol, Aquoll	Gore, 1983	Dimichele et al., 1987
Salt marsh	Fine, in tabular mat		O–A–(Bg)–C	Herbaceous peat or coal, pyrite nodules, oysters and other marine shells	Histosol, Aquent, Aquept	Rabenhorst & Haering, 1989	Retallack, 1990
Carr	Large, in tabular mat		O–A–(Bg)–C	Woody peat or coal, calcareous	Histosol, Aquent, Aquept	Gore, 1983	Retallack & Dilcher, 1988
Fen	Fine, in tabular mat		O–A–(Bg)–C	Herbaceous peat or coal, calcareous	Histosol, Aquent, Aquoll	Gore, 1983	Retallack, 1990
Mangal	Large, in tabular mat		O–A–(Bg)–C	Woody peat or coal, pyritic nodules, oysters or other marine shells	Histosol, Aquent, Aquept	Chapman, 1977	Retallack & Dilcher, 1981

Table 3-3. Paleosol features of different topographic settings.

Topographic position	Root traces	Burrows	Soil structure	Microfabric	Other soil features	Studies of surface soils	Applications to paleosols
Hillslope	Deeply penetrating	Deeply penetrating	Oxidized peds (soil clods) and cutans	Sepic plasmic fabrics	Argillic (Bt) and calcic (Bk) horizons, soil creep of quartz veins or dipping beds, stone lines, colluvial mantles	Ciolkosz et al., 1990; Simon et al., 1990	Williams, 1968; Zbinden et al., 1988; Holland & Zbinden, 1988
Plateau	Deeply penetrating	Deeply penetrating	Oxidized peds and cutans	Sepic plasmic fabrics	Argillic (Bt) and calcic (Bk) horizons, stone lines, truncated profiles, thick laterites and other duricrusts	Goudie, 1973; McFarlane, 1976; Johnson & Watson-Stegner, 1987	Retallack, 1991b
Well-drained lowland	Deeply penetrating	Deeply penetrating	Oxidized peds and cutans	Sepic plasmic fabrics	Argillic (Bt) and calcic (Bk) horizons, mukkarra structure, sand wedges, ice wedges, fossil bones and land snail shells, hematite and other oxidized minerals	Gile et al., 1980; de Wit, 1978	Williams, 1986; Retallack, 1983a, 1986b, 1991b
Intermittently waterlogged lowland	Tabular & deeply penetrating	Mixed terrestrial and aquatic	Oxidized and reduced soil structures	Sepic and asepic microfabrics	Cumulic horizons, salt crusts, valley calcretes, mix of oxidized minerals such as hematite and gley minerals such as siderite	Harden, 1982; Walker & Butler, 1983; Cremeens & Mokma, 1986	Joeckel, 1988; Smith, 1990; Retallack, 1991b
Waterlogged lowland	Tabular mat	Of mainly aquatic creatures	Few relict peds and cutans, bedding	Undulic and other asepic fabrics	Coal, peat, carbonaceous shale, plant fossils, cumulic horizons, siderite or pyrite nodules	Ho & Coleman, 1967; Lytle, 1968	Retallack & Dilcher, 1981; Retallack et al., 1987; Gardner et al., 1988

Table 3-4. Paleosol features favored by different kinds of parent materials.

Parent material	Grain size	Minerals	pH	Soil types	Studies of surface soils	Application to paleosols
Till	Clayey, rocky	Smectite, calcite	Alkaline	Inceptisol, Mollisol, Alfisol	Birkeland, 1984	Retallack, 1980
Loess	Clayey, silty	Smectite, calcite	Alkaline	Inceptisol, Mollisol, Alfisol	Fehrenbacher et al., 1986	Retallack, 1980
Volcanic sand	Clayey, sandy	Smectite, magnetite	Alkaline	Inceptisol, Alfisol	Dethier, 1988	Retallack, 1983a, 1991b
Quartz sand	Sandy	Quartz, hematite	Acidic	Spodosol, Psamment, Dystrochrept	Thompson & Bowman, 1984; Schwartz, 1988	Batten, 1973; Retallack, 1977; Percival, 1986
Alluvium	Clayey	Smectite, kaolinite	Neutral	Inceptisol, Alfisol, Aridisol	Lepsch & Buol, 1974; Walker & Butler, 1983; Busacca & Singer, 1989	Feakes & Retallack, 1988
Marine shale	Clayey	Smectite, illite	Neutral	Inceptisol, Alfisol, Vertisol	Aandahl, 1978	Retallack, 1983a; Holland & Beukes, 1990
Schist	Clayey	Illite, chlorite	Neutral	Alfisol, Ultisol	England & Perkins, 1959; Marron & Popenoe, 1986	Holland, 1984
Limestone	Clayey, rocky	Calcite, kaolinite	Neutral	Entisol, Inceptisol, Oxisol	Ahmad & Jones, 1969; Scholten & Andriesse, 1986	James & Choquette, 1987; Retallack, 1991b
Granitic rocks	Sandy	Quartz	Acidic	Spodosol, Ultisol, Oxisol	Dixon & Young, 1981; Rutherford, 1987	Grandstaff et al., 1986
Basaltic rocks	Clayey	Smectite, magnetite	Neutral	Inceptisol, Alfisol, Vertisol	Lepsch & Buol, 1974; Craig & Loughnan, 1964	Holland et al., 1989; Retallack, 1991b
Ultramatic rocks	Rocky	Serpentine, pyroxene	Neutral	Entisol, Inceptisol	Garcia et al., 1974; Alexander, 1988	Williams, 1968
Volcanic ash	Clayey	Allophane, halloysite	Neutral	Andisol	Neall, 1977; Tan, 1984	Retallack, 1983a, 1991b

Table 3-5. Paleosol features related to time for formation.

Features	Degree of development					Studies of surface soils	Paleosols
	Very weak 10^2 yr	Weak 10^3 yr	Moderate 10^4 yr	Strong 10^5 yr	Very strong 10^6 yr		
Subsurface calcareous (Bk) horizon morphology	None	Wisps	Nodules	Layer	Layer with pisolites, laminae	Gile et al., 1966; Hay & Reeder, 1978; Machette, 1985	Leeder, 1975; Retallack, 1991b
Subsurface clayey (Bt) horizon morphology	Relict bedding or other rock structure	Some clay skins	Clay skins, peds (soil clods), sepic fabric	Pervasive sepic microfabric within peds	Thick (>2 mm) well-structured, omnisepic	Harden, 1982; Birkeland, 1990	Retallack, 1983a, 1986a, 1990
Surface peaty (O) horizon thickness	0-4 cm	4-40 cm	40 cm-4 m	4-40 m	>40 m	Moore & Bellamy, 1973; Falini, 1965	Retallack, 1990
Weathering rind thickness (mm)							
on basalt	0	<0.05	0-1.4	0.6-2	>2	Colman, 1986	--
on andesite	0	<0.1	0-1.2	0.8-2	>2	Colman, 1986	--
on granite	0	<0.2	0.1-1.7	0.3-3	>3	Birkeland, 1984; Hall & Michaud, 1988	--
Grain morphology of pyroxene of hornblende	Pitted Fresh	Etched Pitted	Very etched Pitted	Gone Etched	Gone Very etched	Birkeland, 1984 Hall & Michaud, 1988	Retallack, 1991b --
of quartz	Fresh	Fresh	Fresh	Pitted	Etched	Cleary & Conolly, 1971	--
Harden index (SDI)	0-5	3-12	8-120	20-250	40-250	Harden, 1982, 1990	

paleosols are disfigured by three widespread alterations that occur very early during burial; decompositional loss of soil organic matter (Stevenson, 1969), formation of drab-colored gleyed haloes around remnant organic matter (de Villiers, 1965; Allen, 1986b) and dehydration reddening of ferric hydroxides (Simonson, 1941; Ruhe, 1969). These alterations may transform a gray/brown soil to a gaudy green/red mottled paleosol with minimal amounts of organic matter (Retallack, 1991a). As a consequence, such soil features as degree of reddening and amounts of organic matter and soil N cannot easily be used to interpret paleoenvironment of paleosols, even though they correlate with climatic variables in surface soils (Jenny, 1941; Birkeland, 1984).

The degree of clayeyness of soils and their depth of weathering also are favorites for the development of climofunctions from modern soils, and these are variables that can be measured in paleosols. In this case, however, their application to paleosols is defeated by the unknown relative contributions of climatic temperature and rainfall, and of time for formation, all of which conspire to create clayey and deeply weathered soils.

ALTERNATIVES TO THE FACTOR-FUNCTION APPROACH

Taxonomic Uniformitarianism

One alternative to the factor-function approach that has proven useful in paleoenvironmental interpretation of paleosols is an approach similar to that widely known among paleontologists as taxonomic uniformitarianism (Dodd & Stanton, 1990). If, for example, fossil bones are identified as those of a large fossil alligator or a fossil snail is identified as that of a large helminthoglyptid, then assuming that the fossil creatures had ecological tolerances similar to their living relatives, a warm, moist, frost-free paleoclimate is indicated. Similarly, identification of a paleosol within a modern soil taxonomy may be taken to imply past conditions similar to those enjoyed by such soils today. In my own research of this kind (Retallack, 1983a, 1991b), I have simultaneously tried to employ several different soil classifications, such as those of the U.S. Soil Conservation Service (Soil Survey Staff, 1975), the Food and Agriculture Organization of UNESCO (U.N.FAO, 1971-1981), the South African Department of Agricultural Technical Services (MacVicar et al., 1977), and the Australian CSIRO (Stace et al., 1968). The approach also has been strengthened by the simultaneous comparison of genetically related suites of paleosols, such as the individual rock units of Fig. 3-4. These can be compared as a landscape unit to soil mapping units in local soil surveys or regional soil maps such as those of U.N.FAO (1971-1981). Taxonomic study of paleosols is a very effective way of locating analogous modern profiles and landscapes within the vast descriptive literature on soils for more detailed comparison with specific paleosols. It also is a useful check on inferences based on the factor-function approach.

It can be argued that paleosols should not be identified in classifications of surface soils, which were not designed for this kind of use (Fastovsky & McSweeney, 1989) and are fundamentally different from biological classifications (Fastovsky, 1991). However, these do not seem to be the most serious problems associated with taxonomic uniformitarian interpretation of paleosols. Classification of paleosols should be a better guide to paleoenvironments than biological classification of fossils, because soil taxonomy is based primarily on environmentally significant features. Organisms on the other hand, are classified by a mix of characteristics, some apparently are adaptations to environment, and others primarily reflecting evolutionary ancestry. As a paleontologist by training, I am quite comfortable with the identification of fossil alligators and snails, even though soft part anatomy would be necessary for a zoologist to feel comfortable with an identification. The identification of paleosols in soil classifications demands comparable assumptions that some soil scientists may find uncomfortable. For example, Aridisols cannot be identified among paleosols from pre-existing paleoclimatic data, because that would make circular the interpretation of paleoclimate from them. In view of the relationships here presented (Figs. 3-1, 3-2) a suitable paleopedological definition of Aridisols would include paleosols with carbonate nodules at a reconstructed depth of less than about 1 m. Such a focus on soil features, rather than climatic variables, would eliminate a persistent element of circular reasoning remaining in "Soil Taxonomy" itself (Soil Survey Staff, 1975). A start has been made in modifying the criteria of soil classification to criteria that can reasonably be expected in paleosols (Retallack, 1988b, 1990), and a consensus view of reasonable criteria will emerge, just as paleontologists have developed for fossil identification.

Although soil classifications were designed in part for agricultural and other human uses, they have fundamental underpinnings in genetic concepts from the accumulated experience of several generations of scientists that have thought very deeply about soils (Buol et al., 1989). My use of a variety of soil classifications has impressed on me their similarity in stressing such features as degree of development, and the importance of such materials as peat, carbonate and clay (Retallack, 1990). They do organize data in a way that is useful for trying to understand a paleosol.

The main problem with taxonomic study of paleosols is not these philosophical difficulties but the practical result that the paleoenvironmental implications of taxonomic study are imprecise and difficult to quantify. Even units deep within the taxonomic hierarchy of soil classification or of landscape assemblages of soils commonly have wide environmental range. If the classification is followed out to find specific surface soils analogous with a paleosol, they are seldom identical in all respects (Retallack, 1991b). Each soil and landscape has endured an individual history.

Another problem is the jargon of soil classification. Just as there was resistance among soil scientists to the introduction of "Soil Taxonomy" (Soil Survey Staff, 1975), some geologists feel that it is an unreasonable imposition to have to master the language of soil science (Fastovsky, 1991). Some familiarity with these terms is very useful for publishable research in paleopedology, but for teaching the uninitiated it may be necessary to abridge this formidable lexicon and use equivalent common English terms (some are suggested by Retallack, 1990).

Numerical Models

There has long been a conceptual framework for mathematical modeling of soil formation. Soils can be viewed as energy transformers, that is a body of material changed by the continuing efforts of natural processes (Runge, 1973). They also can be envisaged as open systems to the extent that they represent a boundary between earth and air through which materials move and are transformed (Simonson, 1978). Recent numerical models of the formation of carbonate horizons within soils provide examples of both these flux (Machette, 1985) and process models (McFadden & Tinsley, 1985; Mayer et al., 1988). These particular models have not to my knowledge been applied to buried paleosols, although there is potential to do so. Other models have been usefully employed in interpreting paleoenvironment from paleosols, such as one specifying atmospheric control of C isotopic systematics of paleosol carbonate (Cerling et al., 1989; Quade et al., 1989; Cerling, 1991; McFadden et al., 1991) and another estimating atmospheric oxidation from chemical weathering of paleosol silicate minerals (Holland, 1984; Holland & Zbinden, 1988; Pinto & Holland, 1988; Zbinden et al., 1988; Holland et al., 1989; Holland & Beukes, 1990). Research continues on the conceptual framework of such models (Johnson & Watson-Stegner, 1987; Johnson et al., 1987; Anderson, 1988; Johnson, 1990) and several excellent textbooks also are available on numerical modeling of soil formation (Richter, 1987, 1990; Ross, 1989). Their application is bound to become more widespread as personal computers proliferate.

The principal problem with numerical modeling of paleosols is the assumptions on which they are based. The two examples already mentioned (Cerling, 1991; Holland, 1984) are typical in requiring data on such conditions as original soil porosity, moisture content and biologically respired CO_2. Such information may not be obtained directly from paleosols. Reasonable values and limits can be applied from modern soils to force the models to perform, but how these values resonate through complex equations modeling a paleosol is not always straightforward. Many other models for modern soil formation are rate models, of the general form $\delta x/\delta t$, where x is some measured feature of the soil, and t is the time in years for its formation. Unfortunately, time for formation can only be estimated to an order of magnitude for paleosols (Table 3-5), and such wide error limits are spread even wider from the denominator of complex equations. In contrast, Jenny's (1941) formulation of time as an independent variable obviates this problem so that it is well suited to paleopedological studies, although Jenny's work has been criticized on this basis (Yaalon, 1975). Because of their assumptions, the results of modeling can only be cautiously accepted. Both the models of Cerling (1991) and Holland (1984) have failed in application to specific paleosols because some assumptions were not tenable (Retallack,

1986b, 1992b). This problem is likely to become more severe as models evolve into animated computer games, and the seductive delight of playing obscures the underlying assumptions.

A second problem for the widespread use of numerical modeling is the advanced mathematical and computer skills that they currently demand. Gaining a working understanding of even such simple models as those of Cerling (1991) and Holland (1984) requires an intellectual effort at least comparable to that of learning the language of soil taxonomy. This is not an effort many soil scientists and geologists are going to make until the models become more user friendly. It will also restrict the teaching of numerical modeling to advanced university classes.

SUMMARY

After 50 yr, the factor-function approach to the study of soils is still thriving, and as indicated here, finding new applications to the paleoenvironmental interpretation of paleosols. The relationship between mean annual rainfall and depth to carbonate nodules in soils popularized in Hans Jenny's (1941) book is a good example of a climofunction that can be used to interpret paleoclimate from calcareous paleosols. A new compilation offered here has shown that this is a surprisingly robust and widespread relationship within aridland soils.

Ongoing research on a variety of other relationships between environmental variables and soil features is providing a vast fund of other information of use for the paleoenvironmental interpretation of paleosols. The principal limitation on the use of the factor-function approach to paleosols are the profound alteration of some paleoenvironmentally sensitive features of soils after their burial and the control of some soil features by several environmental variables.

Alternatives to the factor-function approach include identifying a paleosol within a soil classification in order to infer a paleoenvironment similar to modern analogs, or numerically simulating soil-forming processes of a paleosol. Paleosol identification commonly lacks precision and numerical models often founder on assumptions. Both approaches require a considerable effort to master either terminology or mathematics. Much of our understanding of soil classification and of numerical modeling of soils derives from studies of soil-forming factors, so the factor-function approach is unlikely to be supplanted by these other approaches in the foreseeable future.

ACKNOWLEDGMENTS

This work has been possible because of NSF funding, most recently from Grant EAR-9103178. I also am indebted to Marcelo Zarate for access to Argentine soil literature, and to Les McFadden and David Fastovsky for careful review of an earlier draft of this paper.

APPENDIX 3-1

Data for relation between mean annual rainfall and depth of calcic horizon in format: source, rainfall (mm), and depth of carbonate (cm).

1,400,-85	7,115,-17	8,615,-106	8,505,-62	17,315,-43	28,995,-190
1,250,-30	7,117,-22	8,755,-106	8,520,-62	17,259,-35	28,953,-245
1,220,-25	7,240,-22	8,895,-106	8,570,-62	17,305,-45	28,680,-60
1,180,-12	7,130,-24	8,900,-110	8,585,-62	17,298,-17	28,600,-20
1,100,-35	7,180,-24	8,635,-120	8,625,-62	17,298,-20	29,700,-100
2,380,-25	7,245,-24	8,732,-120	8,665,-62	17,240,-60	29,700,-70
2,410,-30	7,210,-25	8,840,-125	8,770,-62	17,305,-25	29,365,-30
2,635,-90	7,360,-27	8,855,-125	8,560,-71	17,259,-35	29,160,-5
2,430,-20	7,180,-33	8,765,-142	8,355,-77	17,305,-30	30,635,-120
2,430,-90	7,120,-35	8,840,-145	8,515,-77	17,298,-83	31,890,-180
2,660,-60	7,210,-35	8,980,-149	8,600,-77	18,257,-44	31,814,-330
2,660,-30	7,270,-38	8,715,-152	8,625,-77	18,276,-30	31,807,-210
2,685,-97	7,165,-41	8,745,-152	8,690,-77	18,257,-60	31,682,-175
2,635,-127	7,280,-43	8,875,-152	8,570,-81	18,276,-50	31,693,-140
2,535,-90	7,455,-43	8,895,-152	8,597,-81	18,276,-56	31,594,-80
2,610,-51	7,260,-51	8,910,-152	8,690,-81	18,276,-60	31,559,-30
2,510,-90	7,240,-53	8,380,0	8,490,-84	18,276,-48	31,519,-40
2,510,-40	7,260,-53	8,410,0	8,707,-84	18,257,-46	31,429,-10
2,305,-42	7,530,-53	8,370,-22	8,720,-84	19,380,-30	32,750,-70
2,305,-60	7,315,-60	8,380,-22	8,760,-84	19,440,-45	32,750,-60
2,330,-40	7,380,-67	8,395,-22	8,355,-92	19,610,-85	32,750,-76
2,380,-70	7,430,-83	8,385,-25	8,540,-92	19,620,-80	33,82,-21
2,510,-90	7,530,-97	8,375,-25	8,585,-92	19,580,-30	33,178,-27
2,510,-80	7,435,-118	8,390,-27	8,625,-92	19,580,-50	33,178,-33
3,650,-85	8,762,-153	8,420,-30	9,500,-110	19,580,-45	33,178,-44
3,820,-134	8,785,-153	8,680,-30	10,350,-65	20,188,0	33,180,-46
3,775,-106	8,990,-153	8,400,-31	10,350,-55	20,45,0	33,200,-41
3,675,-94	8,730,-154	8,410,-31	11,430,-97	20,188,0	33,200,-58
3,300,-76	8,770,-167	8,455,-31	11,325,-35	21,127,-6	33,225,-46
3,350,-44	8,820,-182	8,472,-31	11,97,-34	22,333,-71	33,225,-49
4,400,-81	8,850,-182	8,540,-31	12,664,-191	22,333,-66	33,280,-36
5,520,-70	8,870,-182	8,560,-34	12,730,-270	23,377,-31	33,280,-48
5,530,-75	8,980,-182	8,425,-38	13,500,-147	23,377,-38	33,310,-64
5,550,-60	8,920,-213	8,410,-40	14,400,-40	23,455,-51	33,310,-69
5,770,-110	8,985,-243	8,445,-40	14,400,-130	23,455,-43	33,333,-67
5,770,-80	8,920,-254	8,460,-40	14,400,-110	23,377,-46	33,420,-72
5,710,-95	8,950,-275	8,580,-40	15,483,-60	23,377,-18	34,600,-70
5,685,-114	8,895,-298	8,475,-43	15,483,-95	23,377,-23	35,730,-114
5,650,-70	8,910,-300	8,510,-45	15,500,-105	23,377,-38	35,730,-95
5,650,-80	8,960,-300	8,390,-48	15,213,-23	23,455,-19	35,750,-99
5,770,-120	8,640,-92	8,405,-48	15,177,-28	24,333,-61	35,830,-107
6,760,-60	8,655,-92	8,460,-48	15,177,-25	24,607,-51	36,580,-59
6,760,-81	8,670,-92	8,540,-48	15,177,-60	24,420,-38	37,325,-45
6,760,-100	8,705,-92	8,550,-48	15,177,-41	25,546,-97	38,476,-64
6,760,-161	8,730,-92	8,430,-49	16,513,-40	25,419,-91	38,476,-45
6,760,-60	8,660,-94	8,570,-49	16,443,-76	26,399,-46	38,476,-67
6,750,-58	8,690,-94	8,440,-52	16,570,-76	26,399,-23	38,476,-80
6,730,-80	8,775,-94	8,670,-52	16,389,-91	26,399,-54	38,476,-67
6,730,-87	8,820,-94	8,410,-62	16,435,-76	27,706,-92	38,176,-26
6,620,-50	8,850,-95	8,425,-62	17,298,-80	27,522,-45	39,183,-15
6,620,-30	8,765,-102	8,440,-62	17,298,-80	27,427,-50	39,178,-20
7,110,-12	8,795,-102	8,465,-62	17,305,-63	27,526,-40	39,183,-24
7,140,-13	8,730,-105	8,480,-62	17,305,-30	27,425,-70	39,170,-24

(continued on next page)

APPENDIX 3-1. Continued.

39,173,-40	39,170,-15	40,189,-3	41,480,-44	41,940,-187	41,770,-120
39,180,-18	39,173,-45	40,324,-38	41,360,-65	41,918,-135	41,820,-75
39,183,-14	39,180,-12	40,184,-40	41,290,-50	41,715,-57	41,750,-110
39,195,-32	39,174,-15	40,91,-22	41,330,-50	41,747,-52	41,830,-125
39,193,-19	39,176,-20	41,705,-77	41,290,-58	41,740,-65	41,725,-100
39,185,-28	39,176,-20	41,745,-60	41,320,-50	41,680,-46	41,330,-60
39,176,-10	39,190,-10	41,580,-45	41,320,-34	41,770,-85	41,400,-60
39,176,-22	40,300,-115	41,490,-72	41,320,-52	41,890,-136	41,750,-85
39,183,-25	40,225,-22	41,380,-45	41,330,-45	41,815,-100	41,720,-75
39,183,-23	40,200,-15	41,490,-60	41,953,-154	41,770,-92	
39,183,-31	40,105,-18	41,380,-33	41,950,-168	41,770,-98	

† Sources: 1 = Dan & Yaalon, 1982; 2 = Stace et al., 1968; 3 = Murthy et al., 1982; 4 = Sehgal et al., 1968; 5 = de Wit, 1978; 6 = Jager, 1982; 7 = Arkley, 1963; 8 = Jenny & Leonard, 1935; 9 = U.N.FAO, 1971; 10 = U.N.FAO, 1981; 11 = U.N.FAO, 1977; 12 = Sidhu et al., 1977; 13 = Ahmad et al., 1977; 14 = Courty & Féderoff, 1985; 15 = Glinka, 1931; 16 = Dokuchaev, 1883; 17 = Fedorin, 1960; 18 = Nogina, 1976; 19 = Thorp, 1936; 20 = Campbell & Claridge, 1987; 21 = Tedrow, 1970, 1977; 22 = McCraw, 1964; 23 = Raeside & Cutler, 1966; 24 = Soil Bureau Staff, 1968; 25 = Leamy & Sanders, 1967; 26 = Orbell, 1974; 27 = del Villar, 1957; 28 = Soil Correlation Committee for South America, 1967; 29 = Al Taie et al., 1969; 30 = Bronger & Heinkele, 1989; 31 = Ruhe, 1984; 32 = Hussain et al., 1984; 33 = Marion et al., 1985; 34 = Fadda, 1968; 35 = Bhargava et al., 1981; 36 = Vinayak et al., 1981; 37 = Bal & Buursink, 1976; 38 = Siderius, 1973; 39 = Mulders, 1969; 40 = Dan et al., 1981; 41 = Plaza & Moscatelli, 1989.

REFERENCES

Aandahl, A.R. 1982. Soils of the Great Plains: Land use, crops and grasses. Univ. Nebraska Press, Lincoln.

Ahmad, M., J. Ryan, and R.C. Paeth. 1977. Soil development as a function of time in the Punjab river plains of Pakistan. Proc. Soil. Sci. Soc. Am. 41:1162–1166.

Ahmad, N., and R.L. Jones. 1969. Genesis, chemical properties and mineralogy of limestone-derived soils, Barbados, West Indies. Trop. Agric. 46:1–15.

Alexander, E.B. 1988. Morphology, fertility and classification of productive soils on serpentinized peridotite in California (U.S.A.). Geoderma 41:337–351.

Allen, J.R.L. 1986a. Pedogenic calcretes in the Old Red Sandstone facies (Late Silurian-Early Carboniferous) of the Anglo Welsh area, southern Britain. p. 56–86. *In* V.P. Wright (ed.) Paleosols: Their recognition and interpretation. Blackwells, Oxford, England.

Allen, J.R.L. 1986b. Time scales of colour change in late Flandrian intertidal muddy sediments of the Severn Estuary. Proc. Geol. Assoc. 97:23–28.

Alt, E. 1932. Klimakunde von Mittel- und Südeuropa. Vol. 3(M). p. 288. *In* W. Köppen and R. Geiger (ed.) Handbuch der Klimatologie. Gebruder Borntraeger, Berlin.

Al Taie, F.H., D. Sys, and G. Stoops. 1969. Soil groups of Iraq: Their classification and characterization. Pedologie, Ghent 19:65–148.

Anderson, D.W. 1988. The effect of parent material and soil development on nutrient cycling in temperate ecosystems. Biogeochemistry 5:71–97.

Angevine, C.L., and K.M. Flanagan. 1987. Buoyant sub-surface loading of the lithosphere in the Great Plains foreland basin. Nature (London) 327:137–139.

Arkley, R.J. 1963. Calculation of carbonate and water movement in soil from climatic data. Soil Sci. 96:239–248.

Bal, L., and J. Buursink. 1976. An inceptisol formed on calcareous loess on the "Dast-i-Esan Top" plain in North Afghanistan. Neth. J. Agric. 24:17–24.

Baldwin, B. and C.O. Butler. 1985. Compaction curves. Bull. Am. Assoc. Petrol. Geol. 69:622–629.

Barshad, I. 1966. The effect of variation in precipitation on the nature of clay mineral formation in soils from acid and basic igneous rocks. p. 169–173. *In* L. Heller and A. Weiss (ed.) Proc. Int. Clay Conf. 10 to 24 Jun. Israel Program Sci. Transl., Jerusalem.

Batten, D.J. 1973. Palynology of early Cretaceous soil beds and associated strata. Palaeontology 16:399-424.

Beeunas, M.A., and L.P. Knauth. 1985. Preserved stable isotopic signature of subaerial diagenesis in the 1.2 b.y. Mescal Limestone, central Arizona: Implications for the timing and development of a terrestrial plant cover. Bull. Geol. Soc. Amer. 96:737-745.

Bhargava, G.P., D.K. Pal, B.S. Kapoor, and S.C. Goswami. 1981. Characteristics and genesis of some sodic soils in Indo-Gangetic alluvial plains of Haryana and Uttar Pradesh. J. Indian Soc. Soil Sci. 21:61-70.

Birkeland, P.W. 1984. Soils and geomorphology. Oxford Univ. Press, New York.

Birkeland, P.W. 1990. Soil-geomorphic research—a selective overview. Geomorphology 3:207-224.

Bronger, A., and T. Heinkele. 1989. Micromorphology and genesis of paleosols in the Luochan loess section, China: Pedostratigraphic and environmental implications. Geoderma 45:123-143.

Brook, G.A., M.E. Folkoff, and E.O. Box. 1983. A world model of soil carbon dioxide. Earth Surf. Processes Landforms 8:79-86.

Brown, J., and A.K. Veum. 1974. Soil properties of the International Tundra Biome Project sites. p. 27-38. In A.J. Holding et al. (eds.) Soil organisms and decomposition in tundra. Int. Biome Project, Stockholm, Sweden.

Buol, S.W., F.D. Hold, and R.D. McCracken. 1989. Soil genesis and classification. 3rd ed. Iowa State Univ. Press, Ames, IA.

Busacca, A.J., and M.J. Singer. 1989. Pedogenesis of a chronosequence in the Sacramento Valley, California, U.S.A. II. Elemental chemistry of silt fractions. Geoderma 44:43-75.

Campbell, I.B., and G.G.C. Claridge. 1987. Antarctica: Soils, weathering processes and environment. Elsevier, Amsterdam.

Carlisle, D. 1983. Concentration of uranium and vanadium in calcretes and gypcretes. p. 185-195. In R.C.L. Wilson (ed.) Residual deposits: Surface related weathering processes and materials. Blackwells, Oxford, England.

Cerling, T.E. 1991. Carbon dioxide in the atmosphere: Evidence from Cenozoic and Mesozoic paleosols. Am. J. Sci. 291:377-400.

Cerling, T.E., J. Quade, Y. Wang, and J.R. Bowman. 1989. Carbon isotopes in soils and paleosols as paleoccologic indicators. Nature (London) 341:138-189.

Chapman, V.J. (ed.) 1977. Wet coastal ecosystems. Elsevier, Amsterdam.

Ciolkosz, E.J., B.J. Carter, M.T. Hoover, R.C. Cronce, W.J. Waltman, and R.R. Dobos. 1990. Genesis of soils and landscapes in the Ridge and Valley province of central Pennsylvania. Geomorphology 3:245-261.

Cleary, W.J., and J.R. Conolly. 1971. Distribution and genesis of quartz in a piedmont-coastal plain environment. Bull. Geol. Soc. Am. 82:2755-2766.

Colman, S.M. 1986. Levels of time information in weathering measurements, with examples from weathering rinds on volcanic clasts in the western United States. p. 379-393. In S.M. Colman and D.P. Dethier (ed.). Rates of chemical weathering of rocks and minerals. Academic Press, Orlando, FL.

Courty, M.A., and N. Féderoff. 1985. Micromorphology of recent and buried soils in a semiarid region of northwestern India. Geoderma 35:287-332.

Craig, D.C., and F.C. Loughnan. 1964. Chemical and mineralogical transformation accompanying the weathering of basic volcanic rocks from New South Wales. Aust. J. Soil Res. 2:218-234.

Cremeens, D.L., and D.L. Mokma. 1986. Argillic horizon expression and classification in soils of two Michigan hydrosequences. Soil Sci. Soc. Am. J. 50:1002-1007.

Dan, J., R. Gerson, H. Koyumdjisky, and D.H. Yaalon (ed.). 1981. Aridic soils of Israel. Spec. Publ. 190. Agric. Res. Organization, Bet Dagan, Israel.

Dan, J., and D.H. Yaalon. 1982. Automorphic soils in Israel. p. 103-115. In D.H. Yaalon (ed.) Aridic soils and geomorphic processes. Catena Suppl. 1. Jena, Germany.

Davis, J.C. 1973. Statistics and data analysis in geology. John Wiley & Sons, New York.

de Villiers, J.M. 1965. The genesis of some Natal soils. II. Estcourt, Avalon, Bellevue and Renspruit Series. S. Afr. J. Agric. Sci. 8:507-524.

de Wit, H.A. 1978. Soils and grassland types of the Serengeti Plain (Tanzania). Centre for Agric. Publ. and Documentation, Wangeningen, the Netherlands.

del Villar, E.H. 1957. Soils of the Lusitano-Iberian Peninsula. Translated by G.W. Robinson. Thomas Murby, London.

Dethier, D.P. 1988. Soil chronosequence along the Cowlitz River, Washington. Bull. U.S. Geol. Surv. 1590-F. Reston, VA.

di Castri, F., D.W. Goodall, and R.L. Specht. (ed.). 1981. Mediterranean-type shrublands. Elsevier, Amsterdam.

Dimichele, W.A., T.L. Phillips, and R.G. Olmstead. 1987. Opportunistic evolution, abiotic environmental stress and the fossil record of plants. Rev. Palaeobot. Palynol. 50:151-178.

Dixon, J.C., and R.C. Young. 1981. Character and origin of deep arenaceous weathering mantles on the Bega Batholith, southwestern Australia. Catena 8:97-109.

Dodd, J.R., and R.J. Stanton. 1990. Paleoecology: Concepts and applications. 2nd ed. John Wiley & Sons, New York.

Dokuchaev, V.V. 1883. Russian chernozem. Translated by N. Kaner. Israel Program for Scientific Translation, Jerusalem, Israel.

Emry, R.J., P.R. Bjork, and L.S. Russell. 1987. The Chadronian, Orellan, Whitneyan land-mammal ages, p. 118-152. In M.O. Woodburne (ed.) Cenozoic vertebrates of North America. Univ. California Press, Berkeley.

England, C.B., and H.F. Perkins. 1959. Characteristics of three Reddish Brown Lateritic soils of Georgia. Soil Sci. 88:294-302.

Evanoff, E., D.R. Prothero, and R.H. Lander. 1992. Eocene-Oligocene paleoclimatic change in North America: The White River Formation near Douglas, east-central Wyoming. p. 116-130. In D.R. Prothero and W.A. Berggren (ed.) Eocene-Oligocene climatic and biotic evolution. Princeton Univ. Press, Princeton.

Fadda, G.S. 1968. Étude d'une séquence climatique dans la province de Tucuman (Argentine). Pedologie 18:301-321.

Falini, F. 1965. On the formation of coal deposits of lacustrine origin. Bull. Geol. Soc. Am. 76:1317-1346.

Fastovsky, D.E. 1991. Book review of "Soils of the past." Sedimentology 38:181-184.

Fastovsky, D.E., and K. McSweeney. 1989. Diagenesis, pedogenesis and paleoenvironments: Opening remarks for symposium. Geol. Soc. Am. Abstr. 21:13.

Feakes, C.R., and G.J. Retallack. 1988. Recognition and characterization of fossil soils developed on alluvium: A Late Ordovician example. p. 35-48. In J. Reinhardt and W.R. Sigleo (ed.) Paleosols and weathering through geologic time: Principles and applications. Spec. Pap. 216. Geol. Soc. Am., Boulder, CO.

Fedorin, Y.V. 1960. Soils of the Kazakh S.S.R. 1. The north Kazakhstan region. Translated by A. Gourevitch. Inst. of Soil Sci., Acad. of Sci., Kazakh S.S.R, Alma Ata.

Fehrenbacher, J.B., I.J. Jansen, and K.R. Olson. 1986. Loess thickness and its effects on soils in Illinois. Bull. 782 Illinois Agric. Exp. Stn. Urbana-Champaign, IL.

Folk, R.L., and E.F. McBride. 1976. The Caballos Novaculite revisited. Part 1. Origin of the novaculite members. J. Sediment. Petrol. 46:659-669.

Folk, R.L., H.H. Roberts and C.H. Moore. 1973. Black phytokarst from Hell, Cayman Islands, British West Indies. Bull. Geol. Soc. Am. 84:2351-2360.

Friedmann, E.T., Y. Lipkin, and R. Ocampo-Paus. 1967. Desert algae of the Negev (Israel). Phycologia 5:185-200.

Friedmann, E.I., and R. Weed. 1987. Microbial trace fossil formation, biogenous and abiotic weathering in the Antarctic cold desert. Science (Washington, DC) 236:703-705.

Garcia, A., J. Aguilar, and M. Delgado. 1974. Micromorphological study of soils developed on serpentinite rock from Sierra de Caratraca (Malaga, Spain). p. 394-407. In G.K. Rutherford (ed.) Soil micromorphology. Limestone Press, Kingston, Ontario, Canada.

Gardner, T.W., E.G. Williams, and P.W. Holbrook. 1988. Pedogenesis of some Pennsylvanian underclays. p. 81-101. In J. Reinhardt and W.R. Sigleo (ed.) Paleosols and weathering through geologic time: Principles and applications. Spec. Pap. 216. Geol. Soc. Am., Boulder, CO.

Gile, L.H., J.W. Hawley, and R.B. Grossman. 1980. Soils and geomorphology in the Basin and Range area of southern New Mexico. Guidebook to the Desert Project. Mem. 39. New Mexico Bur. Mines Min. Res., Las Cruces, NM.

Gile, L.H., F.F. Peterson, and R.B. Grossman. 1966. Morphological and genetic sequences of carbonate accumulation in desert soils. Soil Sci. 101:347-360.

Glinka, K.D. 1931. Treatise on soil science. Translated by A. Gourevitch. Israel program for Sci. Trans., Jerusalem.

Gore, A.J.P. (ed.) 1983. Mires—swamp, bog, fen and moor. Elsevier, Amsterdam.

Goudie, A. 1973. Duricrust in tropical and subtropical landscapes. Clarendon Press, Oxford, England.

Grandstaff, D.E., M.J. Edelman, R.W. Foster, E. Zbinden, and M.M. Kimberley. 1986. Chemistry and mineralogy of Precambrian paleosols at the base of the Dominion and Pongola Groups. Precambrian Res. 32:91-131.

Hall, R.D., and D. Michaud. 1988. The use of hornblende etching, clast weathering and soils to date alpine glacial and periglacial deposits: A study from southwestern Montana. Bull. Geol. Soc. Am. 100:458-467.

Hambrey, M.J., and W.D. Harland. (ed.). 1981. Earth's pre-Pleistocene glacial record. Cambridge Univ. Press, Cambridge, England.

Harden, J.W. 1982. A quantitative index of soil development from field descriptions: Examples from a chronosequence in central California. Geoderma 28:1-28.

Harden, J.W. 1990. Soil development on stable landforms and implciations for landscape studies. Geomorphology 3:391-398.

Harris, T.M. 1957. A Rhaeto-Liassic flora in South Wales. Proc. R. Soc. London B147:289-308.

Hay, R.L., and R.J. Reeder. 1978. Calcretes in Olduval Gorge and the Ndolonya Beds of northern Tanzania. Sedimentology 25:649-673.

Hoffman, J.A.J. 1975. Altas climatico de América del Sur. UNESCO, Paris.

Ho, C., and J. Coleman. 1967. Consolidation and cementation of recent sediments in the Atchafalaya Basin. Bull. Geol. Soc. Am. 80:183-192.

Holland, H.D. 1984. The chemical evolution of the atmosphere and oceans. Princeton Univ. Press, Princeton, NJ.

Holland, H.D., and N. Beukes. 1990. A paleoweathering profile from Griqualand West, South Africa: Evidence for a dramatic rise in atmospheric oxygen between 2.2 and 1.9 b.y. Am. J. Sci. 290A:1-34.

Holland, H.D., C.R. Feakes, and E.A. Zbinden. 1989. The Flin Flon paleosol and the composition of the atmosphere 1.8 BYBP. Am. J. Sci. 289:362-389.

Holland, H.D., and E.A. Zbinden. 1988. Paleosols and the evolution of the atmosphere: Part 1. p. 61-82. *In* A. Lerman and M. Meybeck (ed.) Physical and chemical weathering in geochemical cycles. Kluwer Academic, Dordrecht, the Netherlands.

Hussain, M.S., T.H. Amadi, and M.S. Sulaiman. 1984. Characteristics of soils of a toposequence in northwestern Iraq. Geoderma 33:63-82.

Hutchison, J.H. 1982. Turtle, crocodilian and champsosaur diversity changes in the Cenozoic of the north-central region of the western United States. Palaeogeogr. Palaeoclimatol. Palaeoecol. 37:149-164.

Jager, T.J. 1982. Soils of the Serengeti woodlands. Centre for Agric. Publ. Document., Wageningen, the Netherlands.

James, N.P., and P.W. Choquette. (ed.). 1987. Paleokarst. Springer, New York.

Jennings, J.N. 1985. Karst geomorphology. Blackwell, Oxford, England.

Jenny, H.J. 1941. Factors of soil formation. McGraw-Hill, New York.

Jenny, H.J., and C.D. Leonard. 1935. Functional relationships between soil properties and rainfall. Soil Sci. 38:363-381.

Joeckel, R.M. 1988. Geomorphology of a Pennsylvanian land surface: Pedogenesis in the Rock Lake Shale Member, southeastern Nebraska. J. Sediment. Petrol. 59:469-481.

Johnson, D.L. 1990. Biomantle evolution and the redistribution of earth materials and artifacts. Soil Sci. 149:84-102.

Johnson, D.L., and D. Watson-Stegner. 1987. Evolution model of pedogenesis. Soil Sci. 143:349-366.

Johnson, D.L., D. Watson-Stegner, D.N. Johnson, and R.J. Schaetzl. 1987. Proisotropic and proanistropic processes of pedoturbation. Soil Sci. 143:278-291.

Kaemmerer, M., and J.C. Revel. 1991. Calcium carbonate accumulation in deep strata and calcrete in Quaternary alluvial formation in southern Morocco. Geoderma 48:43-57.

Keller, W.D., J.F. Wescott, and A.O. Bledsoe. 1954. The origin of Missouri fire clays. p. 7-46. *In* A. Swineford and N. Plummer (ed.) Clays and clay minerals. Publ. 327. NAS, Washington, DC.

Lander, R.H. 1990. White River diagenesis. Univ. Illinois, Urbana (Diss. Abstr. 91-24449).

Leamy, M.L., and W.M.H. Sanders. 1967. Soils and land use in the Upper Clutha Valley, Otago. Bull. N.Z. Soil Bur. 28. Lower Hutt, NZ.

Leary, R.L. 1981. Early Pennsylvanian geology and paleobotany of the Rock Island County, Illinois, area. Part 1. Geol. Rep. Investigation 37. Illinois State Museum, Springfield, IL.

Leeder, M.R. 1975. Pedogenic carbonate and flood sediment accretion rates: A quantitative model for alluvial arid-zone lithofacies. Geol. Mag. 112:257-270.

Leith, H. 1975. Modelling the primary productivity of the world. p. 237-263. *In* H. Leith and R.H. Whittaker (ed.) Primary productivity of the biosphere. Springer, New York.

Lepsch, I.F., and S.W. Buol. 1974. Investigations on an Oxisol-Ultisol toposequence—S. Paulo state, Brazil. Soil Sci. Soc. Am. Proc. 38:491-496.

Loope, D.B. 1988. Rhizoliths in ancient eolianites. Sediment. Geol. 56:301-314.

Lydolph, P.E. 1977. Climates of the Soviet Union. *In* H.E. Landsberg (ed.) World Surv. of Climatology. Vol 7. Elsevier, Amsterdam.

Lytle, S.A. 1968. The morphological characteristics and relief relationships of representative soils in Louisiana. Bull. 631. Louisiana Agric. Exp. Stn., Baton Rouge, LA.

Machette, M.N. 1985. Calcic soils of the southwestern United States. p. 10-21. *In* D.L. Weide (ed.) Soils and Quaternary geology of the southwestern United States. Spec. Pap. 203. Geol. Soc. Am., Boulder, CO.

MacVicar, C.N., J.M. de Villiers, R.F. Loxton, E. Verster, J.J.N. Lambrecht, F.R. Merryweather, J. Le Roux, T.H. van Rooyen, and H.J. von M. Harmse. 1977. Soil classification: A binomial system for South Africa. Dep. Agric. Tech. Serv., Pretoria, South Africa.

Mann, A.W., and R.D. Horwitz. 1979. Groundwater calcrete deposits in Australia: Some observations from Western Australia. J. Geol. Soc. Aust. 26:293-303.

Marion, G.M., W.H. Schlesinger, and P.J. Fonteyn. 1985. CALDEP: A regional model for soil formation in southwestern deserts. Soil Sci. 139:468-481.

Marron, D.C., and J.H. Popenoe. 1986. A soil catena on schist in northwestern California. Geoderma 17:307-324.

Martin, J.E. 1983. Composite stratigraphic section of the Tertiary deposits of South Dakota. Dakoterra, Rapid City 2(1):1-8.

Mayer, L., L.D. McFadden, and J.W. Harden. 1988. Distribution of calcium carbonate in desert soils: A model. Geology 16:303-306.

McCraw, J.D. 1964. Soils of Alexandra district. Bull. 24. N.Z. Soil Bur., Lower Hutt, New Zealand.

McFadden, L.D., R.G. Amundson, and O.A. Chadwick. 1991. Numerical modelling, chemical and isotopic studies of carbonate accumulation in soils of arid regions. p. 17-35. *In* B.L. Allen and W.D. Nettleton (ed.) Occurrence, characteristics and genesis of carbonate, gypsum, and silica accumulations. SSSA Spec. Publ. 26. ASA, CSSA, and SSSA, Madison, WI.

McFadden, L.D., and J.C. Tinsley. 1985. Rate and depth of pedogenic carbonate accumulation in soils: formulation and testing of a copmartment model. p. 23-41. *In* D.L. Weide (ed.) Soils and Quaternary geology of the Southeastern United States. Spec. Pap. 203, Geol. Soc. Am., Boulder, CO.

McFarlane, M.J. 1976. Laterite and landscape. Academic Press, New York.

Miller, K.G. 1992. The Late Paleogene isotopic record. p. 160-177. *In* D.R. Prothero and W.A. Berggren (ed.) Eocene-Oligocene climatic and biotic evolution. Princeton Univ. Press, Princeton.

Mokma, D.L., and G.F. Vance. 1989. Forest vegetation and origin of some spodic horizons, Michigan. Geoderma 43:311-324.

Moore, P.D., and D.J. Bellamy. 1973. Peatlands. Springer, New York.

Mulders, M.A. 1969. The arid soils of the Balikh Basin (Syrin). Drukkerij Bronder-Offset, NV, Rotterdam, the Netherlands.

Munn, C.C., G.A. Nielsen, and W.F. Mueggler. 1978. Relationships of soils to mountain and foothill range habitat types and production in western Montana. Soil Sci. Soc. Am. J. 42:435-439.

Murthy, R.S., L.R. Hirekirur, S.B. Deshpande, and B.V. Veneka Rao, (ed.). 1982. Benchmark soils of India. Natl. Bur. Soil Surv. Land Use Planning, Nagpur.

Neall, V.E. 1977. Genesis and weathering of Andosols in Taranaki, New Zealand. Soil Sci. 123:400-408.

Neftel, A., H. Oeschger, J. Schwander, B. Stauffer, and R. Zunbrunn. 1982. Ice core sample measurements give atmospheric CO^2 content during the past 40 000 years. Nature (London) 295:220-223.

Nogina, N.A. 1976. Soils of Transbaikal. Indian Natl. Sci. Document. Center, New Dehli.

Orbell, G.E. 1974. Soils and land use of the mid-Manuherikia Valley, central Otago, New Zealand. Bull. 36. New Zealand Soil. Bur., Lower Hutt, New Zealand.

Paton, T.R. 1974. Origin and terminology for gilgai in Australia. Georderma 11:221-242.

Pazos, M.S. 1990. Some features and processes associated with the caliche under humid climate, Balcarce, Argentina. Pedologie 40:141-154.

Percival, C.J. 1986. Paleosols containing an albic horizon: examples from the Upper Carboniferous of northern England. p. 87-111. *In* V.P. Wright (ed.) Paleosols: Their recognition and interpretation. Blackwells, Oxford, England.

Pinto, J.P., and H.D. Holland. 1988. Paleosols and the evolution of the atmosphere. Part II. p. 21-34. *In* J. Reinhardt and W.R. Sigleo (ed.) Paleosols and weathering through geologic time: Principles and applications. Spec. Pap. 216 Geol. Soc. Am., Boulder, CO.

Plaza, J.C.S.L., and G. Moscatelli. 1989. Mapa de suelos de la provincia de Buenos Aires. Secretaria de Agricultura, Ganaderia y Pesca, Buenos Aires, Argentina.

Quade, J., T.E. Cerling, and J.R. Bowman. 1989. Development of the Asian monsoon revealed by marked ecological shift during latest Miocene in northern Pakistan. Nature (London) 342:163-166.

Rabenhorst, M.C., and K.C. Haering. 1989. Soil micromorphology of a Chesapeake Bay tidal marsh: Implications for sulfur accumulation. Soil Sci. 147:339-347.

Raeside, J.D., and E.J. Cutler. 1966. Soils and related irrigation problems of part of Maniototo Plains, Otago. Bull. 23. New Zealand Soil Bur., Lower Hutt, New Zealand.

Retallack, G.J. 1977. Triassic palaeosols from the upper Narrabeen Group of New South Wales. Part II. Classification and reconstruction. J. Geol. Soc. Aust. 24:19-35.

Retallack, G.J. 1980. Late Carboniferous to Middle Triassic megafossil floras from the Sydney Basin. p. 384-430. *In* C. Herbert and R.J. Helby (ed.) A guide to the Sydney Basin. Bull. 26. Geol. Surv. NSW, Sydney, Australia.

Retallack, G.J. 1982. Paleopedological perspectives on the development of grasslands during the Tertiary. Proc. 3rd North Am. Paleont. Conv. 2:417-421.

Retallack, G.J. 1983a. Late Eocene and Oligocene paleosols from Badlands National Park, South Dakota. Spec. Pap. 193. Geol. Soc. Am., Boulder, CO.

Retallack, G.J. 1983b. A paleopedological approach to the interpretation of terrestrial sedimentary rocks: The mid-Tertiary fossil soils of Badlands National Park, South Dakota. Bull. Geol. Soc. Am. 94:823-840.

Retallack, G.J. 1984a. Completeness of the rock and fossil records: Some estimates using fossil soils. Paleobiology 10:58-78.

Retallack, G.J. 1984b. Trace fossils of burrowing beetles and bees in an Oligocene paleosol, Badlands National Park, South Dakota. J. Paleontol. 58:571-592.

Retallack, G.J. 1986a. Fossil soils as grounds for interpreting long term controls on ancient rivers. J. Sediment. Petrol. 56:1-16.

Retallack, G.J. 1986b. Reappraisal of a 2200 Ma-old paleosol from near Waterval Onder, South Africa. Precambrian Res. 32:195-232.

Retallack, G.J. 1988a. Down to earth approaches to vertebrate paleontology. Palaios 3:335-344.

Retallack, G.J. 1988b. Field recognition of paleosols. p. 1-21. *In* J. Reinhardt and W.R. Sigleo (ed.) Paleosols and weathering through geologic time: Principles and applications. Spec. Pap. 216 Geol. Soc. Am., Boulder, CO.

Retallack, G.J. 1990. Soils of the past: An introduction to paleopedology. Unwin-Hyman, London.

Retallack, G.J. 1991a. Untangling the effects of burial alteration and ancient soil formation. Annu. Rev. Earth Planet. Sci. 19:183-206.

Retallack, G.J. 1991b. Miocene paleosols and ape habitats of Pakistan and Kenyua. Oxford Univ. Press, New York.

Retallack, G.J. 1991c. A field guide to mid-Tertiary paleosols and paleoclimatic changes in the high desert of central Oregon—Part 1. Oregon Geol. 53:51-59.

Retallack, G.J. 1992a. Paleosols and changes in climate and vegetation across the Eocene-Oligocene boundary. p. 382-398. *In* D.R. Prothero and W.R. Berggren (ed.) Eocene-Oligocene climatic and biotic evolution. Princeton Univ. Press, Princeton, NJ.

Retallack, G.J. 1992b. Comment on paleoenvironment of *Kenyapithecus* at Fort Ternan. J. Human Evol. 23:363-389.

Retallack, G.J., and D.L. Dilcher. 1981. Early angiosperm reproduction: *Prisca reynoldsii* gen. et sp. nov. from mid-Cretaceous coastal deposits in Kansas. U.S.A. Palaeontographica B179:103-127.

Retallack, G.J., and D.L. Dilcher. 1988. Reconstructions of selected seed ferns. Ann. Mo. Bot. Gard. 75:1010-1057.

Retallack, G.J., G.D. Leahy, and M.D. Spoon. 1987. Evidence from paleosols for ecosystem changes across the Cretaceous-Tertiary boundary in eastern Montana. Geology 15:1090-1093.

Richter, J. 1987. The soil as a reactor. Translated by R. Anlauf and J. Burrough-Boinisch. Catena, Jena, Germany.

Richter, J. 1990. Models for processes in the soil: Programs and excercises. Catena, Jena, Germany.

Ross, S. 1989. Soil processes: A systematic approach. Routledge, London.

Ruffner, J.A. 1985. Climates of the states. 3rd ed. Gale Res. Co. Detroit, MI.

Ruhe, R.V. 1984. Soil climate system across the prairies in the midwestern U.S.A. Geoderma 34:204–219.

Runge, E.C.A. 1973. Soil development and energy models. Soil Sci. 115:183–193.

Rutherford, G.K. 1987. Pedogenesis of two Ultisols (Red Earth Soils) on granite in Belize, Central America. Geoderma 40:225–236.

Rutherford, M.C. 1982. Woody plant biomass distribution in *Burkea africana* savannas. p. 120–141. *In* B.H. Huntley and B.H. Walker (ed.) Ecology of tropical savannas. Springer, Berlin.

Sanchez, P.A., and S.W. Buol. 1974. Properties of some soils of the upper Amazon Basin of Peru. Soil Sci. Soc. Am. Proc. 38:117–121.

Schaetzl, R.J., and C.J. Sorenson. 1987. The concept of "buried" versus "isolated" paleosols: Examples from northeastern Kansas. Soil Sci. 143:426–435.

Scholle, P.A., and P.R. Schluger. 1979. Aspects of diagenesis. Spec. Publ. 26. Soc. Econ. Paleont. Min., Tulsa, OK.

Scholten, J., and W. Andriesse. 1986. Morphology, genesis and classification of three soils over limestone, Jamaica. Geoderma 39:1–40.

Schultz, C.B., and T.M. Stout. 1980. Ancient soils and climatic cycles in the central Great Plains. Trans. Nebraska Acad. Sci. 8:187–205.

Schumm, S.A. 1975. Erosion in miniature pediments in Badlands National Monument, South Dakota. Bull. Geol. Soc. Am. 73:718–724.

Schwartz, D. 1988. Some podzols on Bateke sands and their origins, Peoples Republic of Congo. Geoderma 43:229–247.

Sehgal, J.L., C. Sys, and D.R. Bhumbla. 1968. A climatic soil sequence from the Thar Desert to the Himalayan Mountains in Punjab (India). Pedologie 18:351–373.

Shea, J.H. 1982. Twelve fallacies of uniformitarianism. Geology 10:449–496.

Sherman, G.D. 1952. The genesis and morphology of the alumina-rich laterite clays. p. 154–161. *In* A.F. Fredericksen (ed.) Problems of clay and laterite genesis. Am. Inst. Mining and Metallurgical Eng., New York.

Siderius, W. 1973. Soil transitions in central east Botswana (Africa). Krips Repro, Meppel, the Netherlands.

Sidhu, P.S., J.L. Sehgal, and N.S. Randhawa. 1977. Elemental distribution and association in some alluvium-derived soils of the Indo-Gangetic Plain of Punjab, India. Pedologie 27:225–235.

Simon, A., M.C. Larsen, and C.R. Hupp. 1990. The role of soil processes in determining mechanisms of slope failure and hillslope development in a humid tropical forest, eastern Puerto Rico. Geomorphology 3:263–286.

Simonson, R.W. 1941. Studies of buried soils formed from till in Iowa. Soil Sci. Soc. Am. Proc. 6:373–381.

Simonson, R.W. 1978. A multiple-process model of soil genesis. p. 1–25. *In* W.C. Mahaney (ed.) Quaternary soils. Geoabstracts, Norwich, England.

Singer, A., and E. Galan. 1984. Palygorskite-sepiolite: Occurrence, genesis, uses. Elsevier, Amsterdam.

Smith, R.M.H. 1990. Alluvial paleosols and pedofacies sequences in the Permian Lower Beaufort of the southwestern Karroo Basin, South Africa. J. Sediment. Petrol. 60:258–276.

Soil Bureau Staff. 1968. General survey of the soils of the South Island, New Zealand. Bull. 27. New Zealand Soil Bur., Lower Hutt, NZ.

Soil Correlation Committee for South America. 1967. Report of the meeting of the Soil Correlation Committee for South America. Soil Res. Rep. 30. U.N.FAO, Rome.

Soil Survey Staff. 1975. Soil taxonomy: A basic system of soil classification for making and interpreting soil surveys. USDA-SCS Agric. Handb. 436. U.S. Gov. Print. Office, Washington, DC.

Soil Survey Staff. 1990. Keys to soil taxonomy. Soil Manage. Support Serv. Tech. Monogr. 19. Blacksburg, VA.

Stace, H.C.T., G.D. Hubble, R. Brewer, K.H. Northcote, J.R. Sleeman, M.J. Mulcahy, and H.G. Hallsworth. 1968. A handbook of Australian soils. Rellim, Adelaide, Australia.

Stevenson, F.J. 1969. Pedohumus: Accumulation and diagenesis during the Quaternary. Soil Sci. 107:470–479.

Stoops, G. 1983. Micromorphology of oxic horizons. p. 419–440. *In* P. Bullock and C.P. Murphy (ed.) Soil micromorphology. A.B. Acad., Berkhamsted, England.

Surdam, R.C., and L.J. Crossey. 1987. Integrated diagenetic modeling: A process-oriented approach for clastic systems. Annu. Rev. Earth Planet. Sci. 15:141–170.

Swinehart, J.B., V.L. Souders, H.M. de Graw, and R.F. Diffendal. 1985. Cenozoic paleogeography of western Nebraska. p. 209–229. *In* R.M. Flores and S.S. Kaplan (ed.) Cenozoic paleogeography of the western United States. Rocky Mountain Sect. of Soc. of Econ. Paleontologists and Mineralogists, Denver, CO.

Swisher, C.C., and D.R. Prothero. 1990. Single crystal $^{40}Ar/^{39}Ar$ dating of the Eocene-Oligocene transition in North America. Science (Washington, DC) 249:760–762.

Taha, M.F., S.A. Harb, M.K. Nagib, and A.H. Tantawy. 1981. The climate of the near East. p. 183–255. *In* H.E. Landsberg (ed.) Vol. 9. World Surv. of Climatology. Elsevier, Amsterdam.

Tan, J.H. (ed.) 1984. Andosols. Van Nostrand Reinhold, New York.

Tedrow, J.C.F. 1970. Soil investigations in Inglefield Land, Greenland. Meddel. øm Grønland 188. Copenhagen, Denmark.

Tedrow, J.C.F. 1977. Soils of the polar landscapes. Rutgers Univ. Press, New Brunswick, NJ.

Thompson, C.H., and G.M. Bowman. 1984. Subaerial denudation and weathering of coastal dunes in eastern Australia. p. 263–290. *In* B.G. Thom (ed.) Coastal geomorphology in Australia. Academic Press, Sydney, Australia.

Thorp, J. 1936. Geography of the soils of China. Natl. Geol. Surv. of China, Nanking, China.

Tissot, B.T., and D.H. Welte. 1984. Petroleum formation and occurrence. Springer, Berlin.

United Nations FAO. 1971–1981. Soil map of the world 1:5 000 000. 10 vols. UNESCO, Paris.

United Nations FAO. 1971. Soil map of the world. Vol. 4. South America. UNESCO, Paris.

United Nations FAO. 1977. Soil map of the world. Vol. 7. South Asia. UNESCO, Paris.

United Nations FAO. 1981. Soil map of the world. Vol. 5. Europe. UNESCO, Paris.

van Donselaar-ten Bokkel Huinink, W.A.E. 1966. Structure, root systems and periodicity of savanna plants and vegetation in northern Surinam. North-Holland, Amsterdam.

Viles, H.A. 1987. Blue green algae and terrestrial limestone weathering on Aldabra atoll: An SEM and light microscope study. Earth Surf. Processes Landforms 12:319–330.

Vinayak, A., J.L. Sehgal, and P.K. Sharma. 1981. Pedogenesis and taxonomy of some alluvium-derived sodic soils of Punjab. J. Indian Soc. Soil Sci. 29:71–80.

Walker, B.D., and T.W. Peters. 1977. Soils of Truelove Lowland and Plateau. p. 31–62. *In* L.C. Bliss (ed.) Truelove Lowland, Devon Island: A high Arctic ecosystem. Univ. Alberta Press, Edmonton, Canada.

Walker, P.H., and B.E. Butler. 1983. Fluvial processes. p. 83–90. *In* Div. Soils. CSIRO (ed.) Soils: An Australian perspective. Academic Press, London.

Wanless, H.R. 1923. The stratigraphy of the White River Beds of South Dakota. Am. Philos. Soc. Proc. 62:663–669.

Washburn, A.L. 1980. Geocryology. John Wiley and Sons, New York.

Watts, I.E.M. 1969. Climates of China and Korea. p. 1–117. *In* H.E. Landsberg (ed.) World survey of climatology. Vol. 8. Elsevier, Amsterdam.

Watts, N.L. 1976. Paleopedogenic palygorskite from the basal Permo-Triassic of northwest Scotland. Am. Mineral. 61:299–302.

West, I.M. 1975. Evaporites and associated sediments of the basal Purbeck Formation (Upper Jurassic) of Dorset. Proc. Geol. Assoc. 86:205–225.

Williams, G.E. 1968. Torridonian weathering and its bearing on Torridonian paleoclimate and source. Scott. J. Geol. 4:164–184.

Williams, G.E. 1986. Precambrian permafrost horizons as indicators of paleoclimate. Precambrian Res. 32:233–242.

Wright, V.P. 1981. The recognition and interpretation of paleokarsts: Two examples from the Lower Carboniferous of South Wales. J. Sediment. Petrol. 52:83–94.

Wright, V.P., and M.E. Tucker. 1991. Calcretes: An introduction. p. 1–34. *In* V.P. Wright and M.E. Tucker (ed.) Calcretes. Blackwell, Oxford, England.

Yaalon, D.H. 1975. Conceptual models in pedogenesis. Can soil-forming functions be solved? Geoderma 14:189–205.

Yaalon, D.H. 1983. Climate, time and soil development. p. 233–251. *In* L.P. Wilding et al. (ed.). Pedogenesis and soil taxonomy: Concepts and interactions. Elsevier, Amsterdam.

Zbinden, E.A., H.D. Holland, C.R. Feakes, and S.K. Dobos. 1988. The Sturgeon Falls paleosol and the composition of the atmosphere 1.1 Ga BP. Precambrian Res. 41:141–183.

4 The "State Factor" Approach in Geoarchaeology[1]

Vance T. Holliday
University of Wisconsin
Madison, Wisconsin

ABSTRACT

The study of soils has long been an important component of geoarchaeology (the application of geosciences to archaeological problems). The widest applications of soil science have involved soil chemistry (for detecting the presence, nature, and intensity of human occupation) and the identification of soils as stratigraphic markers and their use as paleoenvironmental indicators. The "state factor" approach to pedology significantly increases the potential applications of soil studies in archaeological contexts. Chronosequences are useful in dating and correlating sites and for predicting the occurrence of sites of a given age. Consideration of the time factor also can profoundly influence interpretations of occupation zones in buried soils. Toposequences and lithosequences can be important in understanding and interpreting environmental change in an archaeological site and, along with biosequences, are useful in (i) reconstructing the relationship of human occupations to paleolandscapes and landscape evolution (ii) reconstructing paleoenvironments. Understanding and interpretation of soil stratigraphy in archaeological contexts also can be greatly enhanced by consideration of the state factors.

Most archaeologists recognize that a relationship exists between the cultural remains they find in the ground and soils. Beyond that simple relationship, however, archaeologists' understanding of what soils can and cannot tell them and indeed, what a soil is and is not, varies tremendously. In general it seems that the applications of soil studies to archaeology are either very large scale, such as the capability of regional soils to support agriculture or use of soils as stratigraphic markers; or very small scale, for example studying the particle-size distribution or chemistry of a soil. There is a significant middle ground in soil studies that is often overlooked, however. This realm of soil science is pedology, which involves investigation of soils as three-dimensional bodies intimately related to the landscape, focusing on their classification and

---
[1] This chapter is a modified version of Holliday (1990).

Copyright © 1994 Soil Science Society of America, 677 S. Segoe Rd., Madison, WI 53711, USA. *Factors of Soil Formation: A Fiftieth Anniversary Retrospective.* SSSA Special Publication 33.

genesis. That aspect of pedology most directly related to archaeology has a geological (rather than agricultural) basis, evolving from Quaternary geology and geomorphology and sometimes referred to as soil-geomorphology (e.g., Ruhe, 1983; Birkeland, 1984; Catt, 1986). Soil-geomorphology is a subdiscipline of pedology with roots in Jenny's (1941) "state factor" approach to soil genesis (Johnson & Hole, 1994). This chapter reviews current and potential applications of pedology and soil-geomorphology in North American archaeology from the standpoint of the state factors.

There is a considerable and growing body of literature concerning the use of soils in archaeological investigations. Much of the initial, substantive work in this area was done in Great Britain (e.g., Cornwall, 1958, 1960), establishing a tradition that continues to thrive (e.g., Limbrey, 1975; Shackley, 1981; Macphail, 1987). In North American archaeology soils were originally used primarily as stratigraphic markers and continue to be so used, with considerable success (e.g., Judson, 1953; Haynes, 1968, 1975; Hoyer, 1980; Reider, 1980, 1982a,b, 1990; Bettis & Thompson, 1982; Ferring, 1982, 1990; Wiant et al., 1983; Styles, 1985; Hajic, 1990). There have been some applications of pedology for reconstructing landscapes and climatic conditions in archaeological contexts at general and site-specific levels (Haynes & Grey, 1965; Reeves & Dormaar, 1972; Thompson & Bettis, 1980; Reider, 1980, 1982a,b, 1990; Blair et al., 1990; Hajic, 1990; Mandel, 1992) and for dating (Foss, 1977; Bischoff et al., 1981; Bettis, 1992). There also are several general discussions of applications of pedology to archaeology (Lotspeich, 1961; Fenwick, 1968; Tamplin, 1969, Rutter, 1978; Olson, 1981; Holliday, 1989a). Finally, studies of soil chemistry, particularly P, have long proven quite useful in indicating the presence and measuring the degree of human occupation (Solecki, 1951; Ahler, 1973; Eidt, 1977, 1985; Woods, 1977; Gordon, 1978; Griffith, 1981; Gurney, 1985; McDowell, 1988).

Butzer (1977), in reviewing Limbrey (1975), comments on the general nature of that volume and absence of a "usable methodology" for soil science in archaeology. This paper is an attempt to begin establishment of such methodologies. A review of the factors of soil formation will be provided followed by a discussion of how a consideration of the various factors can be applied in archaeology. The final discussion will deal with aspects of soil stratigraphy in view of the factors.

FACTORS OF SOIL FORMATION

The state factor approach to soil genesis (Jenny, 1941, 1980) is the theoretical framework for much of pedology. This approach has many applications to archaeology and will be that followed throughout most of this paper. Jenny(1941, 1980) defined the factors of soil formation as climate, organisms (flora and fauna), relief (or landscape setting), parent material, and time, often written as the equation

$$S = f(cl,o,r,p,t\ldots)$$

where the upper case S is the whole soil. This equation defines the state of the soil as a function of the five factors (the state factors) and other, unspecified factors of local or minor importance (...). The equation as a whole has never been solved, but Jenny (1941, 1980) proposed solving the equation by studying the variation in a soil as a function of one factor, keeping the others constant or accounted for. For example, one could study the variation in soils due to differences in climate by keeping all factors except climate constant. Variations in any soil property or properties can then be attributed to variations in climate. This is written

$$S \text{ or } s = f(cl,o,r,p,t)$$

where the lower case s denotes a soil property or properties. Qualitative statements about soils forming as a function of any one factor are called sequences (climosequence, biosequence, toposequence, lithosequence, chronosequence) and quantitative statements, where functions have been solved for any one factor, are called functions (climofunction, biofunction, topofunction, lithofunction, chronofunction).

The state factor approach to the study of soil genesis is not without criticism, which is summarized by Birkeland (1984, p. 162-168). In particular, the factor approach tends to treat the factors individually, although they often act together, such as climate and biota. Additionally, there are other theoretical approaches to soil genesis, such as the energy model of Runge (1973), and some of these are summarized and compared by Gerrard (1981). For the most part, however, the general validity of the state factor approach has been upheld (e.g., Yaalon, 1975; Bockheim, 1980; Birkeland, 1984) and applied in related fields (e.g., Major, 1951). This approach to pedology is particularly useful "from the point of view of a field-oriented geologist-pedologist, working with a wide variety of soils at the earth's surface" (Birkeland, 1984, p. 166). Because this is the same point of view taken by many geoarchaeologists, the state factor approach also is considered valid in archaeological pedology.

THE STATE FACTOR APPROACH TO ARCHAEOLOGICAL PEDOLOGY

The factors of soil formation that are generally of most concern in archaeology are time and climate. Specifically, this includes using soils as indicators of age, past climates and climate change. The following sections will deal with the archaeological applications of the state factor approach to time and climate in pedogenesis. The influence of the other factors will then be discussed. Some of the examples are not related to archaeological research, because so little of this type of soils work has been done in archaeological contexts, but these examples illustrate the principals and the potential for archaeology.

Time and Pedogenesis

The concept that some time must elapse before a soil can form is probably one of the most significant aspects of pedology in archaeology (Holliday, 1992). The presence of a soil in an archaeological site indicates that there has been a significant period of landscape stability, i.e., relatively little or no erosion or deposition. In the author's experience it seems that many investigators assume that in an archaeological site of some depth, especially a stratified site, sedimentation occurred more or less continuously. However, in many situations, such as alluvial or eolian depositional environments, deposition can occur relatively instantaneously; conceivably in a matter of days, certainly in a matter of years or decades. Soils almost invariably take longer to form; usually at least 100 or several hundred years, commonly thousands of years. A case in point is the Lubbock Lake site on the Southern High Plains of Texas, where the writer has conducted geological and pedological studies for several years (Holliday, 1985a,b,c,d,e, 1988b). Sediments ranging from 3 to 6 m thick accumulated episodically over the past 11 000 yr. The periods of deposition and soil formation are well-dated by over 100 radiocarbon ages (Holliday et al., 1983, 1985) and a plot of sedimentation rates through time (Fig. 4-1) shows that the landscape was stable and soils formed for 6 000 out of the past 11 000 yr.

The degree of development of a soil profile or specific pedologic features in a profile can be used as relative indicators of time elapsed after deposition of parent material and, in some situations, as a more or less absolute indicator of age. This application of soils is derived from the concept of the state factors of soil formation. In a situation where there are a number of soils and where the influence of parent material, landscape position, climate, and flora and fauna can be considered negligible, held constant or otherwise accounted for, the soils with stronger profile development can be considered older than those that are less developed. Pedologic features that are time dependent include overall profile morphology as determined by soil indices (Bilzi & Ciolkosz, 1977a; Harden, 1982), profile thickness (Machette, 1975; Birkeland, 1984), illuvial clay content and reddening of the B horizon (Gile et al., 1981; Harden, 1982; Birkeland, 1984; McFadden et al., 1986), calcium carbonate accumulation (Gile et al., 1981; Machette, 1985; McFadden et al., 1986), alteration or formation of certain clay minerals (Shroba & Birkeland, 1983; Birkeland, 1984; McFadden & Hendricks, 1985), and alteration or translocation of certain forms of Fe, Al, and P (W. Scott, 1977; Birkeland et al., 1979; Birkeland, 1984; McFadden et al., 1986).

In an archaeological site with a chronosequence and also producing time-diagnostic artifacts, radiocarbon ages, or some other form of absolute age control, one can determine rates of soil formation and carry this information to other sites in similar situations and use the soils to provide an age for natural or cultural deposits (Fig. 4-2). The Lubbock Lake site is just such a situation. A late-Holocene chronosequence was defined at the site (Holliday, 1985c, 1988a). Rates and characteristic features of soil development were established by combining field and laboratory data with the well-dated geo-

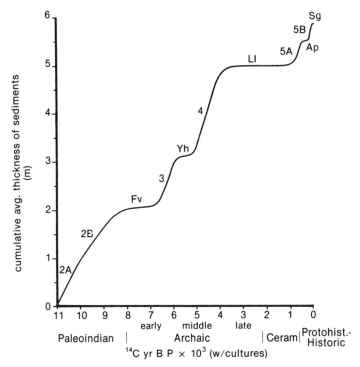

Fig. 4-1. Plot of sedimentation rates at Lubbock Lake over the past 11 000 yr illustrating the episodic nature of depositional events and the relatively long intervals of landscape stability and soil formation (numbers = strata, letters = soils, Fv = firstview soil, Yh = Yellowhouse soil, Ll = Lubbock Lake soil, Ap. = Apache soil, Sg = Singer soil; from Holliday, 1985a).

chronology (Holliday et al., 1983, 1985) (Fig. 4-3, 4-4). The resulting information on pedogenesis is now being used to determine the age of soils, and by inference the age of their parent materials, at other localities in similar settings on the Southern High Plains (e.g., Holliday, 1989b).

In making comparisons of soils from site to site for dating purposes the soils being compared must be in similar landscape positions and parent material. These factors can exert a strong influence on soil morphology even in very young soils (discussed below). Furthermore, other considerations, such as stratigraphic relationships and archaeology must be taken into account. Soils similar in morphology can form at different periods in time.

Baseline studies on rates of soil development such as that described for Lubbock Lake are available or are emerging for other parts of the USA, albeit with varying degrees of age-control reliability. Gile et al. (1981) summarize a classic investigation of soil-geomorphic relations in the desert around Las Cruces, NM, including information on rates of argillic and calcic horizon formation in parent materials of different lithologies and in different landscape positions. G. Scott (1963), Machette (1975) (Fig. 4-2) and Holliday (1987a) discuss various aspects of soil development on terraces of the South

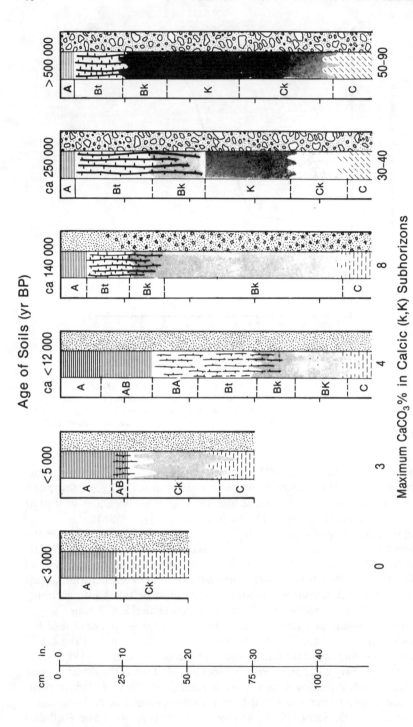

Fig. 4-2. A chronosequence from the South Platte drainage in Colorado showing maximum soil development on parent materials of different ages. The three youngest soils formed in sandy alluvium, the 140 000-yr soil in pebbly sand, and the two oldest soils in sandy gravel (modified from Machette, 1975, with permission).

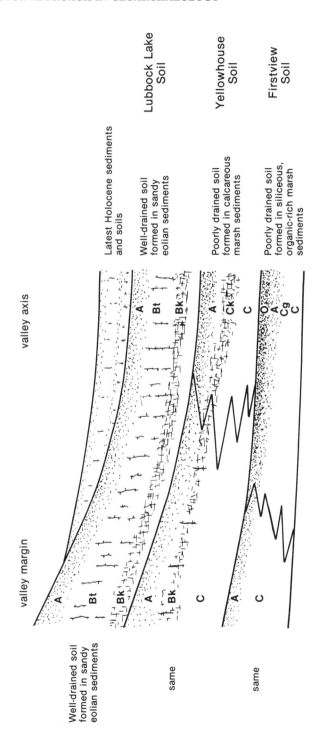

Fig. 4-3. Generalized soil-stratigraphic relationships at the Lubbock Lake site.

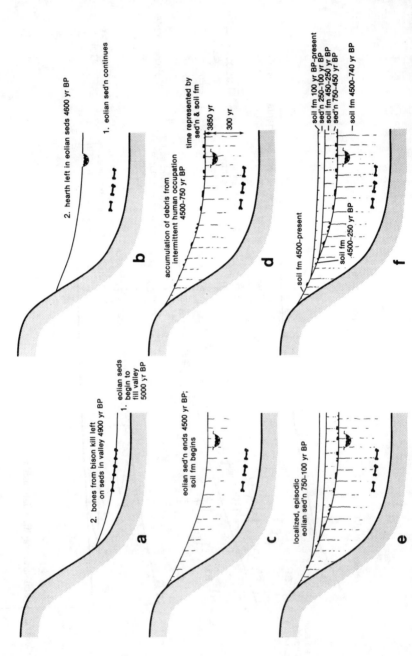

Fig. 4-4. Generalized sequence of selected late-Holocene sedimentological, pedological, and cultural events at the Lubbock Lake site. Numbers 1 and 2 in figures (*a*) and (*b*) indicate sequence of specific events. Figure (*f*) shows present-day soil-stratigraphic and geochronologic relationships along the valley margin. Seds = sediments, sed'n = sedimentation, fm = formation.

Platte River in eastern Colorado, an area famous for its abundance of Paleoindian sites (e.g., Holliday, 1988b). In Wyoming, Reider et al. (1974) established a late Quaternary chronosequence in the Laramie Basin and incorporated their data into related archaeological investigations. A number of chronosequences are available in the far western USA. For example, Shlemon (1978) defined a chronosequence in the southeastern Mojave desert of California and Arizona and has used this data to provide age estimates for several controversial archaeological sites in the area (Bischoff et al., 1978, 1981; Shlemon & Budinger, 1990). McFadden et al. (1986) also carried out research on rates of pedogenesis in the eastern Mojave and McFadden and Weldon (1987) conducted a similar study in the Transverse Ranges of California. Harden et al. (1986) describe a chronosequence along the California coast at Ventura, and Dethier (1988) investigated rates of pedogenesis along the Cowlitz River in western Washington.

Several Holocene chronosequences were investigated in the eastern USA including the upper Susquehanna River basin of New York (Scully & Arnold, 1981), the Ridge and Valley area of central Pennsylvania (Bilzi & Ciolkosz, 1977b), and the Tallapoosa River system in east-central Alabama (Markewich et al., 1988). Limited information on the degree of profile development at archaeological sites in the Brooks Range of Alaska also is available (Reanier, 1982). Finally, Harden and Taylor (1983) present a fine example of the use of soil indices for comparing pedogenesis in chronosequences in different climatic regimes.

In defining various chronosequences and determining rates of pedogenesis, the effects of climate cannot be held constant because the climate has fluctuated considerably during the Quaternary. This does not pose as big a problem as it might seem. A steady rate of soil development does not have to be assumed in using soils as age indicators. For relative age estimates one is simply assigning a qualitative estimate of soil age (e.g., is the soil "young" or old"?). A soil with a well-expressed profile probably had more time to form than one with a poorly expressed profile if both formed in the same area in the same parent material and in similar landscape setting. If degree of soil development is used to make numerical age estimates then one must start with independent age control such as radiocarbon determinations. Then the rate of soil development is calculated relative to either a given number of years or a specified period in the geologic past (e.g., the early Holocene). With sufficient comparative data it should be possible to sort out in a general manner what, if any, effect climate changes had on soil formation. These data can then be applied accordingly when using the soil information to determine ages for other sites in the region.

Climate and Pedogenesis

Often, one of the goals of archaeological research is the reconstruction of the climate when a site or region was occupied or the detection of climate change over a long period of occupation or relative to abandonment. Certain soil properties and soil types are related to climate and if the effects of the other factors of soil formation can be held constant or considered negligible then soils may be able to provide some paleoclimatic data. For a variety of reasons, however, soils probably have limited applications for reconstructing climates in archaeological investigations, the suggestions of Retallack (1990, 1994) notwithstanding. In general, soils are not sensitive to discrete climate changes which may be culturally significant. Such changes can be more easily detected from plant or animal remains. Furthermore, climate changes in the Holocene, the time-period that most North American archaeologists deal with, were often of insufficient magnitude to be detectable in the pedologic record. Finally, soil properties related to the time factor are often difficult to separate from those related to climate. Some pedologic properties, such as reddening of the B horizon, can be time- or climate-dependent. Following a climate change some time is required for the pedologic properties of the soil to reflect the new climate and, in any event, therefore, the comments of Valentine and Dalrymple (1976, p. 218) are well taken, "Although soil science is under great pressure to furnish environmental evidence, it is debatable whether we understand the interaction of the soil-forming processes with the site and environmental factors well enough yet to make confident extrapolations."

In some situations, however, soils have proven useful in providing some general paleoclimatic and paleoenvironmental data. General discussions of the applications of pedologic information to the reconstruction of Quaternary climates are provided by Ruhe (1970), Yaalon (1971), Birkeland (1984), and Retallack (1990). Climate most directly influences pedogenesis through precipitation and temperature (and indirectly through vegetation, discussed below; Birkeland, 1984, p. 295). The properties of soils that seem to best reflect those climatic parameters include overall profile morphology, organic-matter content, and the depth at leaching, which would affect the presence or absence of $CaCO$, and more soluble salts and depth to the top of the zone of accumulation of the carbonate or salts (e.g., Gile et al., 1981; McFadden et al., 1986). Isotopic studies of soil carbonate also are beginning to provide insights into more direct soil–climate relationships (e.g., Cerling, 1984; Cerling & Hay, 1986; Amundson et al., 1989).

Climate change appears to be the dominant factor in producing striking differences in the morphologies of early and middle Holocene soils vs. those of the late Holocene at the Lubbock Lake site (Holliday, 1985b,c) (Fig. 4-4). From 8500 to about 6300 years before the present (YBP) the Firstview Soil developed. It has dark colors suggesting that it contained abundant organic matter, a zone of reduction or gleying immediately below the surface horizon which indicates a water table just below the surface, and, locally, abundant silicified plant remains in the surface horizon. This soil was buried by highly calcareous lake sediments. The Yellowhouse Soil formed in the calcareous sediments from before about 5500 YBP to no later than 5000 YBP. This soil has a thick, organic-rich A horizon. The organic-rich nature of both soils, and the gley horizon in the Firstview Soil and absence of leaching of carbonates in the Yellowhouse Soil suggest a high water table was present throughout the early and into the middle Holocene. In contrast, the late Holocene soils have A-Bw-Bt-Bk horizon sequences typical of well-drained soils in semiarid conditions. These data suggest that the water table was dropping by the end of the middle Holocene, probably due to less effective precipitation. The presence of silica (in silicified plant remains) in the Firstview Soil and $CaCO_3$ in the Yellowhouse Soil suggest a geochemical change in the groundwater from the early to middle Holocene, possibly related to increased temperature. The overall stratigraphic and pedologic sequence, therefore, is indicative of a general drying and warming trend through at least the early and middle Holocene. This trend also is apparent in soils at other localities in the region (Haynes, 1975; Holliday, 1985d,e, 1989b).

Reider (1980, 1982a,b, 1990) used pedologic data for paleoenvironmental reconstructions at archaeological sites in Wyoming. The climatic trends deduced from soils at those sites is generally similar to those from the Southern High Plains. Soils of latest Pleistocene and early Holocene age are typified by dark colors and mottling, which indicate relatively high organic matter production under conditions of impeded drainage and a locally high water table. Soils formed later in the Holocene have profiles suggestive of well-drained conditions and zones of accumulation of carbonate and sometimes more soluble salts, suggesting semiarid to arid climate. Reider and Karlstrom (1987) also used pedologic evidence to infer spring activity in the foothills of the Big Horn Mountains of Wyoming during the middle Holocene, which is otherwise characterized by arid conditions in the region.

An excellent example of the use of soils for climatic reconstructions is provided by Sorenson et al. (1971), Sorenson and Knox (1973) and Sorenson (1977). This research was carried out in and near the forest/tundra ecotone in northern Canada. By mapping the distribution of modern forest and tundra soils and comparing the data to the position of present-day airmasses, then mapping the occurrence of buried and altered forest and tundra soils and dating these soils it was possible to reconstruct paleoairmass frequencies and correlate these shifts to climatic changes during the Holocene.

The concept of the soil-forming interval also is important in the use of soils for climate reconstructions. This idea was formulated and expanded upon primarily by Morrison (1967, 1978) and commonly applied in the western USA. The soil-forming intervals were defined as discrete periods of accelerated soil formation related to particular climatic episodes. Classically, the soil-forming intervals were related to warmer and wetter climate. Little soil formation would take place during intervening cooler and drier episodes. This approach was criticized by some (summarized by Birkeland, 1984, p. 330-334), who argue that basically soil formation always occurs when the landscape is stable, although specific aspects of pedogenesis may vary as a function of climate.

The best example of the soil-forming interval in archaeological pedology in North America is the "Altithermal soil," discussed in some detail by Reider (1990). This is a soil profile, typically buried, reported from many Holocene, primarily alluvial, stratigraphic sequences in the central and western USA (e.g., Leopold & Miller, 1954; Malde, 1964; Haynes, 1968; Reider, 1980, 1982a,b, 1990; Albanese, 1982). The soil is characterized by a moderately well-developed Bw or Bt horizon underlain by a distinct zone of $CaCO_3$ accumulation (Bk). The carbonate accumulation in particular is taken to represent pedogenesis under warmer and possibly drier conditions related to the middle Holocene "thermal maximum" or Altithermal. In very few instances, however, is the precise age of the soil known, particularly the age of burial by the overlying sediments. In the absence of such data an alternative hypothesis is that such a soil is not related to a short interval of drier climate, but formed over a longer period of time. For example, at the famous Clovis site (Blackwater Draw Locality 1) Haynes (1975) reported a well-developed soil believed to have formed in the middle Holocene and considered to be related to a short period of warmer and drier climate. Data now shows that at that site and across the Southern High Plains the middle Holocene or Altithermal was characterized was characterized by eolian deposition not soil formation (Holliday, 1985d, 1989). The well-developed soil observed at Clovis and reported at many other sites in the area formed in the late Holocene (Holliday 1985c,d,e, 1989b). Therefore, it may be argued that in at least some situations climate change or climate extremes are represented by depositional or erosional events rather than periods of soil formation.

Organisms, Relief, and Parent Material

The pedologic information based on the influence of the factors of organisms, relief, and parent material has, at present, more restricted applications in archaeological research than data derived from studies of the state-factors time and climate. There are some aspects of the influence of each of these factors that do have some practical applications in archaeology, however. Flora and fauna certainly exert a tremendous influence on the soils in which they form, and they can leave a variety of physical remains (e.g., pollen and phytoliths) in the soil that can be used as clues to reconstructing floral and faunal assemblages, but these studies are the topics of other disciplines. Plants also influence the isotope chemistry of soils, and studies of stable C isotope are yielding valuable insights into past vegetation (Cerling, 1984; Cerling & Hay, 1986; Amundson et al., 1989).

Beyond isotopes, there is only a limited amount of information relating pedologic features to past plant and animal communities. As well, plant and animal distribution is so closely linked to climate that sorting out the effects of each is often difficult. The best results achieved along these lines have been vegetation reconstructions in restricted areas where the regional climate can be assumed to have been constant at any one time. Most of this work focused on studies of the ecotone between forested and unforested (including both prairie and tundra) regions. Alfisols with E horizons and Spodosols will typically form under forested conditions while Mollisols, Alfisols without E horizons, and Inceptisols are common in the nonforested areas. Because the forest boundary fluctuated through time as a function of climate change, the forest soils left their imprint in the soils of the nonforested areas. The above-cited work by Sorenson et al. (1971), Sorenson and Knox (1973) and

Sorenson (1977) is an example of such a study. The mapping and dating of buried and altered forest and tundra soils allowed for a reconstruction of shifts in the forest–tundra boundary throughout the Holocene. A number of similar sorts of investigations have been carried out along the ecotone of the forests of eastern North America and the prairies of the Great Plains (e.g., Ruhe & Cady, 1969; Ruhe, 1970; Al-Barrak & Lewis, 1978, Anderson, 1987) and in an archaeological context at the foot of the Canadian Rockies (Reeves & Dormaar, 1972). Soil morphology and chemistry, along with other lines of evidence, also was used to dispel the long-held belief that the Southern High Plains was covered by a boreal forest at the close of the Pleistocene, the time of the earliest human occupation of the region (Holliday, 1987b).

It is important to understand how the properties of a soil can vary as the parent material or relief changes. These variations can influence soil profile morphology such that tracting a particular soil or using a soil as a stratigraphic marker could be difficult. This could be especially significant when working on a large archaeological site or over a larger area such as a drainage basin and where only limited natural or artificial exposures are available. On the time scale of the Holocene probably one of the most significant influences that parent material can have on a soil is in water movement through the soil. Coarser-textured soils allow much faster through-flow of water than fine-grained soils. Gile et al. (1966, 1981), for example, show quite clearly how the morphology of calcic horizons varies markedly as the texture of the parent material changes. Furthermore, textural variations in parent material at one locality (e.g., an alluvial deposit consisting of a layer of fine sand over a lens of gravelly sand) also can significantly influence the resultant profile morphology.

Clayey soils, particularly Vertisols, can have a profound influence on archaeological sites. The shrinking and swelling characteristic of this type of soil can destroy the stratigraphic and cultural contexts of a site (Duffield, 1970; Wood & Johnson, 1978; Johnson & Watson-Stegner, 1990). Duffield (1970) goes on to suggest that in Texas, where Vertisols are widespread and were probably difficult to till by prehistoric agriculturalists, these soils may have been significant in restricting the westward spread of village farmers.

Slope position can affect soil profile morhpology in several ways. Within a large area such as a drainage basin the orientation of the slopes is important. For example, Lotspeich and Smith (1953) show that significant differences in soil morphology are apparent between north-facing and south-facing slopes. Water movement over and through soils also is strongly influenced by slope position. It has been observed in many situations that soils near the summits of slopes are better drained, but receive less effective moisture (due to run off) than soils at the foot of slopes (e.g., Ruhe, 1969; Daniels et al., 1971; Jenny, 1980; Hall, 1983; Birkeland, 1984). Soils in upslope positions also will tend to have material constantly removed from their surfaces, whereas the soils at the foot of the slope will tend to accumulate the materials eroded from upslope. The morphology of a soil can therefore vary markedly along a slope (e.g., Swanson, 1985; Berry, 1987; Birkeland et al., 1991). An understanding of such pedologic variation as a function of slope position is particularly significant when using soils to date landscapes.

The concept of variation in soil morphology as a function of landscape setting can be quite useful in the reconstruction of landscapes and local environments. This principle is particularly applicable to buried soils, which represent buried landscpaes. Daniels and Jordan (1966) and Ruhe (1969), among others, have applied these principles to regional and local reconstructions of Quaternary landscapes in a series of landmark investigations in Iowa. In strictly archaeological contexts, this principal is well illustrated at the Lubbock Lake site. Holliday (1985b) shows that there is considerable variation in the above-mentioned Firstview Soil as a function of landscape setting and microenvironment. Along the valley margin the soil is well drained, as indicated by oxidation colors (Fig. 4-3). The presence of some $CaCO_3$ in the profile and a thin A horizon with low organic C content also suggests relatively dry conditions in this position. Along the valley axis the soil was very poorly drained (Fig. 4-3), as described above. A somewhat similar situation was described for the overlying Yellowhouse Soil. At the Clovis Site, several early to middle Holocene soils also exhibit considerable variation due to landscape setting (Haynes, 1975; Holliday, 1985d). Archaeological uses of buried soils to reconstruct landscapes are further discussed by Ferring (1990, 1992) and case histories include the Cherokee (Hoyer, 1980), Delaware Canyon (Ferring, 1982), Napoleon Hollow (Wiant et al., 1983; Styles, 1985), and Koster (Hajic, 1990) sites, and studies in western Kansas (Mandel, 1992) and the Missouri drainage of Iowa (Thompson & Bettis, 1980).

SOILS AS STRATIGRAPHIC MARKERS

Soils commonly have been used as stratigraphic markers in archaeological research. The unique physical and chemical properties that distinguish soils from sediments make soils quite useful for stratigraphic subdivision and correlation. However, the nature of soils also necessitates the exercise of a certain amount of caution in their use as stratigraphic markers.

The presence of a soil in a stratigraphic sequence marks the passage of some amount of time with no or very little erosion or sedimentation, but the parent material for that soil may have been deposited virtually instantaneously, as discussed earlier. If the soil in such a sequence is buried, the contact of the top of the soil with the sediments that bury it marks the gap in time. This means that cultural material found at the top and bottom of even a thick deposit may be nearly contemporaneous, whereas the artifacts near the top of the unit may be considerably older than an artifact immediately above, on the paleosurface (Fig. 4-4). Moreover, artifacts found on the paleosurface may well represent a mixture of cultural material left on the surface during the entire period of soil formation (Fig. 4-4). The artifacts and occupations also are subject to mixing by biological and geological activities (Duffield, 1970; Wood & Johnson, 1978; Hole, 1981; Stein, 1983 Johnson & Watson-Stegner, 1990) as well as stratigraphic compression. This situation was encountered at the Lubbock Lake site (Johnson & Holliday, 1986), Wilson-Leonard site, in central Texas (Holliday, 1992), at several sites in southwestern Oklahoma (Ferring, 1982), Shawnee Minisink site, Delaware (Dent, 1985), and probably occurs at many other stratified sites. On the positive side, the surface of a buried soil may be a zone likely to contain archaeological material because it does represent a stable landscape. A consideration of the time represented by the surface of a buried soil also is important in interpreting radiocarbon ages determined on the organic matter found in buried A horizons. Because the organic matter accumulates over time, the radiocarbon age will be the nonmathematical average age or "apparent mean residence time" of the A horizon plus the time since burial. Samples from buried A horizons are subject to contamination by younger organic compounds moving in from overlying soils, among other problems (Campbell et al., 1967; Scharpenseel, 1971, 1979; Martel & Paul, 1974; Burleigh, 1974; Matthews, 1985). Under proper circumstances, however, such dating can be useful in providing a minimum date for deposition of the parent material and beginning of pedogenesis and a maximum date for burial of the soil (e.g., Holliday et al., 1983, 1985; Matthews, 1985; Haas et al., 1986).

In using buried soils as stratigraphic markers it also should not be assumed that the amount of time represented by the contact with the overlying sediment is always the same. Burial of the surface may take place at different times in different places (Fig. 4-4).

Soils can be useful in identifying various strata, but they should never be considered strata themselves or referred to as "strata" or "layers," or otherwise treated like geological deposits. This point is well made by Tamplin (1969). At the Lubbock Lake site almost all of the A horizons of the various buried soils have been designated as some sort of strata since the first excavations in 1939. It cannot be overemphasized that a soil profile is imprinted over geologic deposits through time. Furthermore, the boundaries between soil horizons often have no relationship to geological layering. Depending on the slope of the surface associated with the soil and variations in the permeability of the parent material, soil horizons can crosscut depositional layering. At the Lubbock Lake site a prominent calcic horizon was designated as a "strata" during early investigations and artifacts were excavated accordingly (Johnson & Holliday, 1986). It is now apparent that where the earlier excavations took place the calcic horizon crosscuts two geologic deposits and an intervening buried soil (Fig. 4-4). Artifacts found in the calcic horizon could be as much as 6000 yr old or less than 5000 yr old.

Furthermore, it should be kept in mind that pedogenesis can obscure sedimentary features. One of the criteria often used to differentiate between a B horizon and a C horizon is whether bedding in the parent material has been obscured. Evidence of beddding can persist in a B horizon if the original beds included lenses of gravel; otherwise the sedimentological history of the parent material will have to come from laboratory data. However, the original particle-size distribution of the parent material also can be obscured by pedogenesis. Significant amounts of illuvial (post depositional) clay can accumulate in Holocene soils, which could make some sedimentological interpretations difficult. For example, the B horizon of a soil that had 3500 yr to form at the Lubbock Lake site has as much as 6% illuvial clay (Holliday, 1985c, 1988a).

CONCLUSIONS

Soils have long been recognized as important components of archaeological sites. They have most commonly been used as stratigraphic markers. Soil chemistry also has been used as a tool to identify habitation areas or specific activity areas or to detect evidence of agricultural activity. Soils have been viewed less commonly as natural entities, constituting a type of near-surface weathering phenomena that grows as a three-dimensional body in sediment through time, under the influence of its parent material, the slope of the associated surface, and the climate. In addition, archaeological data can be useful to pedologists (e.g., Sandor et al., 1986a,b,c; Collins & Shaprio, 1987; Foss & Collins, 1987).

The above-outlined approach to soils is derived from pedology, which is concerned with the genesis and classification of soils and founded on strong geomorphic principles and data. In particular, this soil-geomorphic approach to geoarchaeology and pedology is derived largely from the work and ideas of Hans Jenny, beginning with his landmark 1941 volume, which we honor here, and continuing through his 1980 work.

Many archaeologists seem to have some familiarity with soil classification, but in archaeological pedology and soil–geomorphology classification is only one of many research tools and the complete integration of pedology into these areas of research requires an understanding of soils far beyond basic terminology. No archaeologist is expected to become a pedologist any more than they are expected to become palynologists, zoologists, or sedimentologists. However, an understanding of the basic principles of pedology, as well as other disciplines, is essential for communication with specialists in such disciplines. A knowledge of the potentials and limitations offered by soils and knowing what questions to ask of a pedologist can aid considerably in conducting an efficient and more complete investigation.

ACKNOWLEDGMENTS

I thank Ron Amundson for his work in arranging this 50th anniversary perspective and for his considerable efforts in revising the original paper that was the basis for this manuscript.

REFERENCES

Ahler, S.A. 1973. A chemical analysis of deposits at Rogers Rock Shelter, Missouri. Plains Anthropol. 18:116–131.
Albanese, J. 1982. Geologic investigationp. p. 309–330. *In* G.C. Frison and D.J. Stanford (ed.) The Agate Basin site. Acad. Press, New York.

Al-Barrak, S., and D.T. Lewis. 1978. Soils of a grassland-forest ecotone in eastern Nebraska. Soil Sci. Soc. Am. J. 42:334-338.

Amundson, R.G., O.A. Chadwick, J.A. Sowers, and H.E. Doner. 1989. The stable isotope chemistry of pedogenic carbonate at Kyle Canyon, Nevada. Soil Sci. Soc. Am. J. 53:201-210.

Anderson, D.W. 1987. Pedogenesis in the grassland and adjacent forests of the Great Plains. Adv. Soil Sci. 7:53-93.

Berry, M.E. 1987. Morphological and chemical characteristics of soil catenas on Pinedale and Bull Lake moraine slopes in the Salmon River Mountains, Idaho. Quat. Res. 28:210-225.

Bettis, E.A. 1992. Soil morphologic properties and weathering zone characteristics as age indicators in Holocene alluvium in the Upper Midwest. p. 119-144. *In* V.T. Holliday (ed.) Soils and landscape evolution. Smithsonian Inst. Press, Washington, DC.

Bettis, E.A., and D.M. Thompson. 1982. Interrelationships of cultural and fluvial deposits in northwest Iowa. Association of Iowa archaeologists fieldtrip guidebook. Univ. South Dakota Archaeol. Lab., Vermillion, SD.

Bilzi, A.F., and E.J. Ciolkosz. 1977a. A field morphology rating scale for calculating pedogenic development. Soil Sci. 124:45-48.

Bilzi, A.F., and E.J. Ciolkosz. 1977b. Time as a factor in the genesis of four soils developed in recent alluvium in Pennsylvania. Soil Sci. Soc. Am. J. 41:122-127.

Birkeland, P.W. 1984. Soils and geomorphology. Oxford Univ. Press, New York.

Birkeland, P.W., A.L. Walker, J.B. Benedict, and F.B. Fox, 1979. Morphological and chemical trends in soil chronosequences: Alpine and arctic environments. p. 188. *In* Agronomy abstracts. ASA, Madison, WI.

Birkeland, P.W., M.E. Berry, and D.K. Swanson. 1991. Use of soil catena field data for estimating relative ages of moraines. Geology 19:281-283.

Bischoff, J.L., W.M. Childers, and R.J. Shlemon. 1978. Comments on the Pleistocene age assignment and associations of a human burial from the Yuha Desert, California: A rebuttal. Am. Antiq. 43:747-749.

Bischoff, J.L., R.J. Shlemon, T.L. Ku, R.D. Simpson, R.J. Rosenbauer, and F.E. Budinger, 1981. Uranium-series and soil-geomorphic dating of the Calico archaeological site, California. Geology 9:576-582.

Blair, T.C., J.S. Clark, and S.G. Wells. 1990. Quaternary continental stratigraphy, landscape evolution, and application to archaeology: Jarilla piedmont and Tularosa graben floor, White Sands Missile Range, New Mexico. Geol. Soc. Am. Bull. 102:749-759.

Bockheim, J.G. 1980. Solution and use of chronofunctions in studying soil development. Geoderma 24:71-85.

Burleigh, R. 1974. Radiocarbon dating: Some practical considerations for the archaeologist. J. Archaeol. Sci. 1:69-87.

Butzer, K.W. 1977. Review of 'Soil science in archaeology'. Am. Antiq. 42:303-304.

Campbell, C.A., E.A. Paul, D.A. Rennie, and K.J. McCallum. 1967. Factors affecting the accuracy of the carbon-dating method in soil humus studies. Soil Sci. 104:81-85.

Catt, J.A. 1986. Soils and Quaternary geology: A handbook for field scientists. Oxford Sci. Publ., Oxford, England.

Cerling, T.E. 1984. The stable isotopic composition of soil carbonate and its relationship to climate. Earth Planet. Sci. Lett. 71:229-240.

Cerling, T.E., and R.L. Hay. 1986. An isotopic study of paleosol carbonates from Olduvai Gorge. Quat. Res. 25:63-78.

Collins, M.E., and G. Shapiro. 1987. Comparisons of human-influenced and natural soils at the San Luis archaeological site, Florida. Soil Sci. Soc. Am. J. 51:171-176.

Cornwall, I.W. 1958. Soils for the archaeologist. Phoenix House, London.

Cornwall, I.W. 1960. Soil investigations in the service of archaeology. Viking Fund Publ. Anthropol. 28:265-299.

Daniels, R.B., and R.H. Jordan. 1966. Physiographic history and the soils, entrenched stream systems, and gullies, Harrison County, Iowa. USDA Tech. Bull. 1348. U.S. Gov. Print. Office, Washington, DC.

Daniels, R.B., E.E. Gamble, and J.G. Cady. 1971. The relation between geomorphology and soil morphology and genesis. Adv. Agron. 23:51-88.

Dent, R.J. 1985. Amerinds and the environment: Myth, reality, and the upper Delaware valley. p. 123-163. *In* C.W. McNett (ed.) Shawnee Minisink: A stratified Paleoindian-Archaic site in the Upper Delaware Valley of Pennsylvania. Acad. Press, Orlando, FL.

Dethier, D.P., 1988. The soil chronosequence along the Cowlitz River, Washington. U.S. Geol. Surv. Bull. 1590-F. U.S. Gov. Print. Office, Washington, DC.

Duffield, L.F. 1970. Vertisols and their implications for archaeological research. Am. Anthropol. 72:1055–1062.

Eidt, R.C. 1977. Detection and examination of anthrosols by phosphate analysis. Science (Washington, DC) 197:1327–1333.

Eidt, R.C. 1985. Theoretical and practical considerations in the analysis of anthrosols. p. 155–190. *In* G. Rapp and J.A. Gifford (ed.) Archaeological geology. Yale Univ. Press, New Haven, CT.

Fenwick, I.M. 1968. Pedology as tool in archaeological investigations. Ont. Archaeol. Publ. 11:27–38.

Ferring, C.R. 1982. The late Holocene prehistory of Delaware Canyon, Oklahoma. Contributions in archaeology 1. Inst. of Applied Sci., North Texas State Univ., Denton, TX.

Ferring, C.R. 1990. Archaeological geology of the Southern Plains. p. 253–266. *In* N.P. Lasca and J. Donahue (ed.) Archaeological geology of North America. Geol. Soc. Am. Centennial Vol. 4.

Ferring, C.R. 1992. Alluvial pedology and geoarchaeological research. p. 1–40. *In* V.T. Holliday (ed.) Soils in archaeology. Smithsonian Inst. Press, Washington, DC.

Foss, J.E. 1977. The pedological record at several Paleoindian sites in the Northeast. p. 234–244. *In* W.S. Newman and B. Salwen (ed.) Amerinds and their paleoenvironments in northeastern North America. Ann. New York Acad. Sci. 228:234–244.

Foss, J.E., and M.E. Collins. 1987. Future users of soil genesis and morphology in allied science. p. 293–299. *In* L.L. Boersma (ed.) Future developments in soil science research. SSSA, Madison, WI.

Gerrard, A.J. 1981. Soils and landforms. Allen and Unwin, London.

Gile, L.H., F.F. Peterson, and R.B. Grossman. 1966. Morphological and genetic sequences of carbonate accumulation in desert soils. Soil Sci. 101:347–360.

Gile, L.H., J.W. Hawley, and R.B. Grossman. 1981. Soils and geomorphology in the Basin and Range area of southern New Mexico—Guidebook to the Desert Project. New Mexico Bur. of Mines and Mineral Resour., Mem. 39, Socorro, NM.

Gordon, B.C. 1978. Chemical and pedological delimiting of deeply stratified archaeological sites in frozen ground. J. Field Archaeol. 5:331–338.

Griffith, M.A. 1981. A pedological investigation of an archaeological site in Ontario, Canada. II: Use of chemical data to discriminate features of the Benson site. Geoderma 25:27–34.

Gurney, D.A. 1985. Phosphate analysis of soils: A guide for the field archaeologist. Inst. Field Archaeol. Tech. Pap. 3. Inst. Field Archaeol. London.

Haas, H., V.T. Holliday, and R. Stuckenrath. 1986. Dating of Holocene stratigraphy with soluble and insoluble organic fractions at the Lubbock Lake site, Texas: An ideal case study. Radiocarbon 28:473–485.

Hajic, E.R. 1990. Koster site archeology I: Stratigraphy and landscape evolution. Center for Am. Archeology Res. Ser. 8, Kampsville, IL.

Hall, G.F. 1983. Pedology and geomorphology. p. 117–140. *In* L.P. Wilding et al. (ed.) Pedogenesis and soil taxonomy. Part 1. Developments in Soil Sci. 11A. Elsevier, Amsterdam.

Harden, J.W. 1982. A quantitative index of soil development from field descriptions: Examples from a chronosequence in central California. Geoderma 28:1–28.

Harden, J.W., and E.M. Taylor. 1983. A quantitative comparison of soil development in four climatic regimes. Quat. Res. 20:342–359.

Harden, J.W., A.M. Sarna-Wojcicki, and G.R. Dembroff. 1986. Soils developed on coastal and fluvial terraces near Ventura, California. U.S. Geol. Surv. Bull. 1590-B. U.S. Gov. Print. Office, Washington, DC.

Haynes, C.V., Jr. 1968. Geochronology of late-Quaternary alluvium. p. 591–631. *In* R.B. Morrison and H.E. Wright, Jr. (ed.) Means of correlation of Quaternary successions, Univ. Utah Press, Salt Lake City, UT.

Haynes, C.V., Jr. 1975. Pleistocene and recent stratigraphy. p. 58–96. *In* F. Wendorf and J.J. Hester (ed.) Late Pleistocene environments of the Southern High Plains. Ft. Burgwin Res. Center, Publ. 9. Taos, NM.

Haynes, C.V., Jr., and D.C. Grey. 1965. The Sister's Hill site and its bearing on the Wyoming postglacial alluvial chronology. Plains Anthropol. 10:196–207.

Hole, F.D. 1981. Effects of animals on soils. Geoderma 25:72–112.

Holliday, V.T. 1985a. Archaeological geology of the Lubbock Lake site, Southern High Plains of Texas. Geol. Soc. Am. Bull. 96:1483–1492.

Holliday, V.T. 1985b. Early Holocene soils at the Lubbock Lake archaeological site, Texas. Catena 12:61–78.

Holliday, V.T. 1985c. Morphology of late Holocene soils at the Lubbock Lake archaeological site, Texas. Soil Sci. Soc. Am. J. 49:938-946.
Holliday, V.T. 1985d. New data on the stratigraphy and pedology of the Clovis and Plainview sites, southern High Plains. Quat. Res. 23:368-402.
Holliday, V.T. 1985e. Holocene soil-geomorphological relationships in a semi-arid environment: The southern High Plains of Texas. p. 321-353. *In* J. Boardman (ed.) Soils and Quaternary landscape evolution. John Wiley & Sons, Chichester, England.
Holliday, V.T. 1987a. Geoarchaeology and late Quaternary geomorphology of the middle South Platte River, northeastern Colorado. Geoarchaeology 2:317-329.
Holliday, V.T. 1987b. A reexamination of late-Pleistocene boreal forest reconstructions for the Southern High Plains. Quat. Res. 28:238-244.
Holliday, V.T. 1988a. Genesis of a late Holocene soil chronosequence at the Lubbock Lake archaeological site, Texas. Ann. Assoc. Am. Geogr. 78:594-610.
Holliday, V.T. (ed.). 1988b. Guidebook to the archaeological geology of the Colorado Piedmont and High Plains of southeastern Wyoming. Geography Dept., Texas A&M Univ., College Station, TX.
Holliday, V.T. 1989a. Paleopedology in archaeology. p. 187-206. *In* A. Bronger and J. Catt (ed.) Paleopedology: Nature and applications of paleosols. Catena Suppl. 16. Cremlingen, Germany.
Holliday, V.T. 1989b. Middle Holocene drought on the Southern High Plains. Quat. Res. 31;74-82.
Holliday, V.T. 1990. Pedology in archaeology. p. 525-540. *In* N.P. Lasca and J. Donahue (ed.) Archaeological geology of North America. Geol. Soc. Am. Cent. Vol. 4, Boulder, CO.
Holliday, V.T. 1992. Soil formation, time, and archaeology. p. 101-117. *In* V.T. Holliday (ed.) Soils and landscape evolution. Smithsonian Inst. Press, Washington, DC.
Holliday, V.T., E. Johnson, H. Haas, and R. Stuckenrath. 1983. Radiocarbon ages from the Lubbock Lake site, 1950-1980: Framework for cultural and ecological change on the southern High Plains. Plains Anthropol. 28:165-182.
Holliday, V.T., E. Johnson, H. Haas, and R. Stuckenrath. 1985. Radiocarbon ages from the Lubbock Lake site, 1981-1984. Plains Anthropol. 30:277 291.
Hoyer, B.E. 1980. The geology of the Cherokee Sewer site. p. 21-66. *In* D.C. Anderson and H.A. Semken (ed.) The Cherokee excavations. Acad. Press, New York.
Jenny, H. 1941. Factors of soil formation. McGraw-Hill, New York.
Jenny, H. 1980. The soil resource. Springer-Verlag, New York.
Johnson, D.J., and F.D. Hole. 1994. Soil formation theory: A summary of its principal aspects on geography, soil geomorphology, Quaternary geology, and paleopedology. p. 111-126. *In* R. Amundson et al. Factors of soil formation: A fiftieth anniversary retrospective. SSSA Spec. Publ. 33. SSSA, Madison, WI.
Johnson, D.L., and D. Watson-Stegner. 1990. The soil-evolution model as a framework for evaluating pedoturbation in archaeological site formation. p. 541-560. *In* N.P. Lasca and J. Donahue (ed.) Archaeological geology of North America. Geol. Soc. Am. Cent. Vol. 4. Boulder, CO.
Johnson, E., and V.T. Holliday. 1986. The Archaic record at Lubbock Lake. Plains Anthropol. Mem. 21:7-54.
Judson, S. 1953. Geology of the San Jon site, eastern New Mexico. Smithson. Misc. Collect. 121:1-70.
Leopold, L.B., and J.P. Miller. 1954. A post-glacial chronology for some alluvial valleys in Wyoming. U.S. Geol. Surv. Water Supply Pap. 1261. U.S. Gov. Print. Office, Washington, DC.
Limbrey, S. 1975. Soil science in archaeology. Acad. Press, London.
Lotspeich, F.B. 1961. Soil science in the service of archaeology. p. 137-139. *In* F. Wendorf (ed.) Paleoecology of the Llano Estacado. Fort Burgwin Res. Center. Publ. 1. Taos, NM.
Lotspeich, F.B., and H.W. Smith. 1953. Soils of the Palouse loess: I. The Palouse Catena. Soil Sci. 76:467-480.
Machette, M.N. 1975. Geologic map of the Lafayette quadrangle, Adams, Boulder, and Jefferson Counties, Colorado. U.S. Geol. Surv. Map MF-656. U.S. Gov. Print. Office, Washington, DC.
Machette, M.N. 1985. Calcic soils of the southwestern United States. p. 1-21. *In* D.L. Weide (ed.) Soils and Quaternary geology of the Southwestern United States. Geol. Soc. Am. Spec. Pap. 203, Boulder, CO.
Macphail, R.I. 1987. A review of soil science in archaeology in England. p. 332-379. *In* H.C.M. Keeley (ed.) Environmental archaeology: A regional review. English Heritage Occasional Pap. 1, London.

Major, J. 1951. A functional, factorial approach to plant ecology. Ecology 32:392-412.
Malde, H.E. 1964. Environment and man in arid America. Science (Washington, DC) 145:123-129.
Mandel, R.D. 1992. Soils and Holocene landscape evolution in central and southwestern Kansas: Implications for archaeological research. p. 41-100. *In* V.T. Holliday (ed.) Soils and landscape evolution. Smithsonian Inst. Press, Washington, DC.
Markewich, H.W., W.C. Lynn, M.J. Pavich, R.G. Johnson, and J.C. Meetz. 1988. Analyses of four Inceptisols of Holocene age, east-central Alabama. USGS Bulletin 1589-C. U.S. Gov. Print. Office, Washington, DC.
Martel, Y.A., and E.A. Paul. 1974. The use of radiocarbon dating of organic matter in the study of soil genesis. Soil Sci. Soc. Am. J. 38:501-506.
Matthews, J.A. 1985. Radiocarbon dating of surface and buried soils: Principles, problems and prospects. p. 269-288. *In* K.S. Richards et al. (ed.) Geomorphology and soils, Allen and Unwin, London.
McFadden, L.D., and D.M. Hendricks. 1985. Changes in the content and composition of pedogenic iron oxyhydroxides in a chronosequence of soils in southern California. Quat. Res. 23:189-204.
McFadden, L.D., and R.J. Weldon. 1987. Rates and processes of soil development on Quaternary terraces in Cajon Pass, California. Geol. Soc. Am. Bull. 98:280-293.
McFadden, L.D., S.G. Wells, and J.C. Dohrenwend. 1986. Influences of Quaternary climate changes on processes of soil development on desert loess deposits of the Cima Volcanic field, California. Catena 13:361-389.
McDowell, P.F. 1988. Chemical enrichment of soils at archaeological sites: Some Oregon case studies. Phys. Geogr. 9:247-262.
Morrison, R.B. 1967. Principals of Quaternary soil stratigraphy. p. 1-69. *In* R.B. morrison and H.E. Wright, Jr. (ed.) Quaternary soils. Desert Res. Inst., Univ. Nevada, Reno.
Morrison, R.B. 1978. Quaternary soil stratigraphy-concepts, methods, and problems. p. 77-108. *In* W.C. Mahaney (ed.) Quaternary soils. Geogr. Abstracts, Norwich, England.
Olson, G.W. 1981. Soils and the environment: A guide to soil surveys and their applications. Chapman and Hall, New York.
Reanier, R.E. 1982. An application of pedological and palynological techniques at the Mesa site, northern Brooks Range, Alaska. Anthropol. Pap. Univ. Alaska 20:123-191.
Reeves, B.O.K., and J.F. Dormaar. 1972. A partial Holocene pedological and archaeological record for the southern Alberta Rocky Mountains. Arct. Alp. Res. 4:325-336.
Reider, R.G. 1980. Late Pleistocene and Holocene sols of the Carter/Kerr-McGee archeological site, Powder River basin, Wyoming. Catena 7:301-315.
Reider, R.G. 1982a. Soil development and paleoenvironments. p. 331-344. *In* G.C. Frison and D. Stanford (ed.) The Agate Basin site. Acad. Press, New York.
Reider, R.G. 1982b. The soil of Clovis age at the Sheaman archaeological site, eastern Wyoming. Contrib. Geol. 21:195-200.
Reider, R.G. 1990. Late Pleistocene and Holocene pedogenic and environmental trends at archaeological sites in plains and mountain areas of Colorado and Wyoming. p. 335-360. *In* N.P. Lasca and J. Donahue (ed.) Archaeological geology of North America. Geol. Soc. Am. Cent. Vol. 4, Boulder, CO.
Reider, R.G., N.J. Kuniansky, D.M. Stiller, and P.J. Uhl 1974. Preliminary investigation of comparative soil development on Pleistocene and Holocene geomorphic surfaces of the Laramie basin, Wyoming. p. 27-33. *In* M. Wilson (ed.) Applied geology and archaeology: The Holocene history of Wyoming. Rep. Invest. 10. Geol. Surv. Wyoming, Laramie, WY.
Reider, R.G., and E.T. Karlstrom. 1987. Soils and stratigraphy of the Laddie Creek site (48BH345), an Altithermal-age occupation in the Big Horn Mountains, Wyoming. Geoarchaeology 2:29-47.
Retallack, G.J. 1990. Soils of the past: An introduction to paleopedology. Unwin and Hyman, New York.
Retallack, G.J. 1994. The environmental factor approach to the interpretation of paleosols. p. 31-64. *In* R. Amundson et al. (ed.) Factors of soil formation: A fiftieth anniversary retrospective. SSSA Spec. Publ. 33. SSSA, Madison, WI.
Ruhe, R.V. 1969. Quaternary landscapes in Iowa. Iowa State Univ. Press, Ames, IA.
Ruhe, R.V. 1970. Soils, paleosols, and environment. p. 37-52. *In* W. Dort and J.K. Jones, Jr. (ed.) Pleistocene and recent environments of the Central Great Plains. Univ. Kansas Press, Lawrence, KS.

Ruhe, R.V. 1983. Aspects of Holocene pedology in the United States. p. 12–25. *In* H.E. Wright, Jr. (ed.) Late-Quaternary environments of the United States. Vol. 2. Univ. Minnesota Press, Minneapolis.

Ruhe, R.V., and J.G. Cady. 1969. The relation of Pleistocene geology and soils between Bentley and Adair in southwestern Iowa. USDA Tech. Bull. 1349. U.S. Gov. Print. Office, Washington, DC.

Runge, E.C.A. 1973. Soil development sequences and energy models. Soil Sci. 11:518–319.

Rutter, N.W. 1978. Soils in archaeology: Geosciences in Canada, 1977. Ann. Rep. and Rev. of Soil Sci., Geol. Surv. Canada Pap. 78-6. Geol. Surv. of Canada, Ottawa.

Sandor, J.A., P.L. Gersper, and J.W. Hawley. 1986a. Soils at prehistoric agricultural terracing sites in New Mexico: I. Site placement, soil morhpology, and classification. Soil Sci. Soc. Am. J. 50:166–173.

Sandor, J..A, P.L. Gersper, and J.W. Hawley. 1986b. Soils at prehistoric agricultural terracing sites in New Mexico: II. Organic matter and bulk density changes. Soil Sci. Soc. Am. J. 50:173–177.

Sandor, J.A., P.L. Gersper, and J.W. Hawley. 1986c. Soils at prehistoric agricultural terracing sites in New Mexico: III. Phosphorus, selected micronutrients, and pH. Soil Sci. Soc. Am. J. 50:177–180.

Scharpenseel, H.W. 1971. Radiocarbon dating of soils. Sov. Soil Sci. 3:76–83.

Scharpenseel, H.W. 1979. Soil fraction dating. p. 277–283. *In* R. Berger and H.E. Suess (ed.) Radiocarbon dating, Proc. 9th Int. Radiocarbon Conf., Los Angeles and La Holla, 1976. Univ. California Press, Berkeley.

Scott, G.R. 1963. Quaternary geology and geomorphic history of the Kassler quadrangle, Colorado. U.S. Geol. Surv. Prof. Pap. 421-A. U.S. Gov. Print. Office, Washington, DC.

Scott, W.E. 1977. Quaternary glaciation and volcanism, Metolius River area, Oregon. Geol. Soc. Am. Bull. 88:113–124.

Scully, R.W., and R.W. Arnold. 1981. Holocene alluvial stratigraphy in the upper Susquehanna River basin, New York. Quat. Res. 15:327–344.

Shackley, M.L. 1981. Environmental archaeology. Allen and Unwin, London.

Shlemon, R.J. 1978. Quaternary soil-geomorphic relationships, southeastern Mojave desert, California and Arizona. p. 187–207. *In* W.C. Mahaney (ed.) Quaternary soils. Geogr. Abstr., Norwich, England.

Schleman, R.J., and F.E. Budinger. 1990. The archaeological geology of the Calico site, Mojave Desert, California. p. 301–314. *In* N.P. Lasca and J. Donahue (ed.) Archaeological geology of North America. Geol. Soc. Am. Cent. Vol. 4, Boulder, CO.

Shroba, R.R., and P.W. Birkeland. 1983. Trends in late-Quaternary soil development in the Rocky Mountains and Sierra Nevada of the western United States. p. 145–146. *In* H.E. Wright, Jr. (ed.) Late-Quaternary environments of the United States. Vol. 1. Univ. Minnesota Press, Minneapolis.

Solecki, R.S. 1951. Notes on soil analysis and archaeology. Am. Antiq. 16:254–256.

Sorenson, C.J. 1977. Reconstructed Holocene bioclimates. Ann. Assoc. Am. Geogr. 67:214–222.

Sorenson, C.J., and J.C. Knox. 1973. Paleosols and paleoclimates related to late Holocene forest/tundra border migrations: Mackenzie and Keewatin, N.W.T., Canada. p. 187–204. *In* S. Raymons and P. Schlederman (ed.) Int. Conf. on the Prehistory and Paleoecology of Western North American Arctic and Subarctic, Calgary, Canada. November 1972. Univ. Calgary Arch. Assoc., Univ. Alberta Print. Dep., Edmonton.

Sorenson, C.J., K.C. Knox, J.A. Larsen, and R.A. Bryson. 1971. Paleosols and the forest border in Keewatin, N.W.T. Quat. Res. 1:468–473.

Stein, J.K. 1983. Earthworm activity: A source of potential disturbance of archaeological sediments. Am. Antiq. 48:277–289.

Styles, T.R. 1985. Holocene and late Pleistocene geology of the Napoleon Hollow site in the lower Illinois valley. Cent. for Am. Arch., Kampsville Archeol. Cent., Res. Ser. 5, Kampsville, IL.

Swanson, D.K. 1985. Soil catenas on Pinedale and Bull Lake moraines, Willow Lake, Wind River Mountains, Wyoming. Catena 12:329–342.

Tamplin, M.J. 1969. The application of pedology to archaeological research. p. 153–161. *In* S. Pawluk (ed.) Pedology and Quaternary research. Univ. Alberta Print. Dep., Edmonton.

Thompson, D.M., and E.A. Bettis. 1980. Archeology and Holocene landscape evolution in the Missouri drainage of Iowa. J. Iowa Archeol. Soc. 27:1–60.

Valentine, K.W.G., and J.B. Dalrymple. 1976. Quaternary buried paleosols: A critical review. Quat. Res. 6:209–222.

Wiant, M.D., E.R. Hajic, and T.R. Styles. 1983. Napoleon Hollow and Koster site stratigraphy: Implications for Holocene landscape evolution and studies of Archaic period settlement patterns in the lower Illinois valley. p. 147-164. *In* J.L. Phillips and J.A. Brown (ed.) Archaic hunters and gatherers in the American Midwest. Acad. Press, New York.

Wood, W.R., and D.L. Johnson. 1978. A survey of disturbance processes in archaeological site formation. Adv. Archaeol. Method Theory 1:315-381.

Woods, W.I. 1977. The quantitative analysis of soil phosphate. Am. Antiq. 42:248-251.

Yaalon, D.H. 1971. Soil-forming processes in time and space. p. 29-40. *In* D.H. Yaalon (ed.) Paleopedology. Univ. Israel Press, Jerusalem.

Yaalon, D.H. 1975. Conceptual models in pedogenesis: Can soil-forming functions be solved? Geoderma 14:189-205.

5 Factors Controlling Ecosystem Structure and Function

Peter M. Vitousek
Stanford University
Stanford, California

ABSTRACT

Factors of Soil Formation is a seminal book in terrestrial ecosystem ecology much as it is in pedology. The insights and syntheses therein remain a driving force in studies of natural and managed ecosystems. The influence of *Factors of Soil Formation* is illustrated by recent examples of ecological studies based explicitly on the climate, organism, relief, parent material, time, and human activity factors. Where single-factor studies are impractical, ecosystem studies treat the interactions of state factors (with each other and with processes internal to ecosystems) in process-based models whose development and validation are themselves dependent on state factor-based approaches. Finally, the legacy of *Factors of Soil Formation*, and the man who created it, is now being felt in the development of ecological research programs to analyze causes, consequences, and feedbacks of global environmental change.

Factors of Soil Formation is focused on ecosystems—the term is synonymous with the "larger system" that is the explicit focus of much of Jenny's work. Jenny did not invent the ecosystem concept—but *Factors of Soil Formation*, and other publications by Jenny, introduced ways of thinking about ecosystems that today continue to represent a dominant conceptual approach to the field. Jenny's contributions include: (i) he identified potentially independent factors that could control terrestrial ecosystems (in an ultimate sense), and distinguished these from processes internal to ecosystems, and (ii) he suggested, and demonstrated, that analyzing variations in the structure and dynamics of ecosystems in relation to variations in those independent factors yields insight into the control of ecosystem processes—and a crucial background for experimental analyses.

In addition to these conceptual underpinnings, *Factors of Soil Formation* brought a number of empirical approaches to the attention of ecosystem scientists—and synthesized a great deal of basic information that remains useful today. Consequently, I believe it is reasonable to regard Jenny as the

---

Copyright © 1994 Soil Science Society of America, 677 S. Segoe Rd., Madison, WI 53711, USA. *Factors of Soil Formation: A Fiftieth Anniversary Retrospective*. SSSA Special Publication 33.

intellectual founder of terrestrial ecosystem ecology. He is certainly recognized widely within ecology; of approximately 180 citations to *Factors of Soil Formation* recorded in *Science Citation Index* from 1980 to 1990, 92 were in soils or agricultural journals and 43 were in ecological or biological journals (the remainder are not easy to classify). Moreover, this simple list greatly understates his effect on the field; his more recent book (Jenny, 1980) draws many citations each year, and his groundbreaking analysis of the controls of litter decomposition (Jenny et al., 1949) also continues to be cited actively by biologists.

In this paper, I show how each of the major state factors identified by Jenny (1941) continues to provide an organizing framework for ecological studies. I then discuss studies of interactions among factors and their integration in ecosystem models. Finally, I demonstrate the utility of Jenny's approach to developing studies of the causes and consequences of global environmental change.

STATE FACTORS IN ECOSYSTEM ANALYSIS

It would be impossible to review state factor-based analyses in ecosystem ecology—they pervade the field. Instead, I will describe a single example of a recent ecological study based on each of the major factors.

Climate

The climate factor accounts for the largest amount of variation in ecosystem structure and function globally; indeed, temperature and precipitation (independently) probably account for more of that variation that any other single factor. At a time when the scale of human activity is becoming sufficiently large to alter climate regionally and globally, it is particularly important that the influence of this factor be understood.

Ecological studies of variation in ecosystems resulting from variations in climate (climosequences) are carried out on scales ranging from elevation/temperature gradients on individual mountains (cf., Körner, 1989) to global analyses across all of the major biomes (Vogt et al., 1986). As a result of such studies, many of the great patterns that were identified and analyzed by Jenny (i.e., organic matter and N accumulation in high latitude systems) are becoming better understood.

One recent example is a study of vegetation, soils, and nutrient cycling carried out along an elevational gradient of wet tropical forests on Volcan Barva in Costa Rica (Marrs et al., 1988; Heaney & Proctor, 1989). This study contributes towards understanding the pattern and regulation of nitrogen limitation to primary production in terrestrial ecosystems—and it represents a logical outgrowth of Jenny's study of decomposition along an altitudinal gradient in the mountains of Colombia (Jenny et al., 1949). In the Volcan Barva study, Heaney and Proctor (1989) determined litterfall, its chemistry, and the turnover of surface organic matter along a transect on volcanic soils

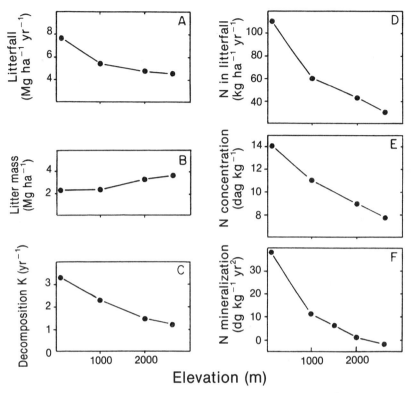

Fig. 5-1. Patterns of mass and N fluxes along an altitudinal gradient on Vólcan Barva, Costa Rica; *A)* Annual leaf litterfall, *B)* Mean mass of leaf litter on the soil surface, *C)* The calculated composition constant (cf., Jenny et al., 1949) for leaf litter decomposition, *D)* Quantity of N in annual litterfall, *E)* Mean concentration of N in litterfall, *F)* Net N mineralization in in situ incubations, *A* through *E* are based on data from Heaney and Proctor (1989, and *F* is based on data from Marrs et al. (1988).

from a lowland tropical forest at 100-m elevation to an upper montane forest at 2600 m; Marrs et al. (1988) measured soil N mineralization along the same gradient (Fig. 5-1). Annual nonwoody litterfall decreased from 7.6 to 4.6 Mg ha^{-1} with increasing altitude, but the mass of litter on the soil surface increased from 2.3 to 3.7 Mg ha^{-1} across the same altitudinal range. More interestingly, N concentrations in litterfall decreased with increasing elevation while P concentrations remained constant, a pattern that also has been observed in other tropical mountains. These results suggest that N availability (in both relative and absolute terms) decreases as temperature decreases in the montane tropics—and measurements of soil N mineralization reinforce this suggestion. Overall, this study supports Grubb's (1977) conclusion that systematic variation in N availability with elevation (= temperature) helps to regulate ecosystem dynamics in tropical montane forests.

Organisms

The biotic factor is a logical topic for ecological study—but as Jenny (1941) made clear, the biotic factor provides conceptual difficulties more severe than those for other factors. Jenny (1980) wrote; "The real bugbear was the biotic factor. Like everybody else, I could see that vegetation affects the soil and that soil affects the vegetation, the very *circulus vitiosus* that I was trying to avoid." Jenny's solution was to identify the regional flora (and fauna), the *potential* occupants of a site, as the factor of interest. This definition avoids the problem of trying to determine cause and effect in the feedback system that characterizes plant-soil interaction.

Recent ecological analyses of the effects of organisms on ecosystem function have used a number of approaches. One approach is based on experimental studies in which replicated monocultures of different species have been planted into an initially homogeneous site, allowing the resulting plant-soil feedbacks to be analyzed directly (Wedin & Tilman, 1990). This approach can link understanding of plant ecophysiology and allocation with that of litter decomposition and soil processes.

Another recent approach has been to examine the ecosystem consequences of alterations in the regional flora—through biological invasions by introduced species, or as a consequence of species extinctions. The ongoing anthropogenic loss of biological diversity, together with the systematic breakdown in biogeographic barriers to dispersal of the remaining species, unfortunately provides an ever-increasing number of opportunities for such studies.

The biota of oceanic islands is disproportionately altered by biological invasion and extinction, and hence such islands are particularly useful for studies of the biotic factor. My colleagues and I have been examining the consequences of invasion by an actinorrhizal N fixer, *Myrica faya*, on soils and nutrient cycling in young volcanic soils in Hawaii (Vitousek et al., 1987; Vitousek & Walker, 1989). *Myrica* was brought to Hawaii from the Canary Islands about 100 yr ago, it is now spreading rapidly into sites which until recently had lacked any symbiotic N fixing species. This alteration of the biotic factor alters ecosystem state and function, as demonstrated by the observations that:

1. Growth of the native vegetation is strongly limited by N availability in the absence of *Myrica*.
2. *Myrica* invasion increases N inputs more than fivefold from background conditions, and causes accumulation of organic C and N in soils.
3. The N fixed by *Myrica* quickly becomes available to other organisms in the soil.
4. The population and activity of other organisms (both plants and soil biota) are influenced by *Myrica* invasion (Aplet, 1990; Walker & Vitousek, 1991) (Fig. 5-2).

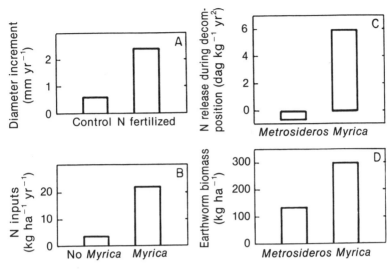

Fig. 5-2. Effects of invasion by the exotic N-fixing tree *M. faya* on ecosystems in Hawaii Volcanoes National Park; *A*) Effects of fertilization with N on growth of the native vegetation, *B*) Total N inputs (including precipitation and N fixation) in a site with heavy *Myrica* colonization vs. one where colonization was prevented, *C*) N release in the 1st yr of decomposition from decaying leaves of *Myrica* vs. the dominant native tree, *Metrosideros polymorpha*, *D*) Earthworm biomass under *Myrica* vs. *Metrosideros*, *A* through *C* are based on data from Vitousek and Walker (1989), and *D* is from Aplet (1990).

Relief

A number of ecological studies have focused on the dynamics of nutrient cycling and loss as they are influenced by soil and water movement along local topographic gradients (cf., Peterjohn & Correll, 1984; Schimel et al., 1985); this approach is one of the fundamental tools of the developing field of landscape ecology. Perhaps the most significant question in landscape ecology, and indeed regional studies in general, is how we can "scale" our understanding of processes at one level of resolution (i.e., plots) to progressively coarser scales (landscapes, continents, ultimately the Earth system). This question was addressed directly in the recent FIFE project [First International Satellite Land Surface Climatology Project (ISLSCP) Field Experiment], a National Aeronautics and Space Administration (NASA)-sponsored study designed to test satellite remote sensing as a tool for integrating information on energy, water, and C exchange across spatial scales from local to regional (Sellers et al., 1988). The study was carried out in and near Konza Prairie, a remnant tall grass prairie in the Flint Hills of Kansas—many of its results are still being analyzed.

In the FIFE study, Schimel et al. (1991) determined patterns of variation in resource availability (light, water, and N) and vegetation response across a number of toposequences. They demonstrated that ecosystem properties varied predictably across these sequences as long as management was reasonably constant. The repeating variation in ecosystems along the toposequences was therefore useful in extrapolating measurements made at a few sites to the estimation of fluxes over the region as a whole.

Parent Material

One recent example of an ecological study of the influence of parent material is the ecosystem gradient on Blackhawk Island, WI. The island supports a continuous gradient of soil texture from sand to clay loam soils, and associated vegetation from pine (*Pinus resinosa*) to sugar maple (*Acer saccharum*) (Pastor et al., 1982). The overall gradient has been used to examine the relationships between N availability, N cycling and primary productivity in forest ecosystems (Pastor et al., 1984), to study root production as a function of soil fertility (Aber et al., 1985), and to evaluate remote sensing as a tool for the determination of forest canopy biogeochemistry (Wessman et al., 1988). It also has been used in the development and parameterization of a widely cited model of forest ecosystem dynamics (Pastor & Post, 1986).

The results of research on Blackhawk Island suggest that N availability is the proximate factor controlling vegetation composition and productivity. However, N availability isn't an independent factor—rather it is determined by soil–plant interaction on both short and long time scales. The texture of soil parent material is the ultimate controlling factor; its effect on N availability is mediated through the effects of soil texture on moisture availability and of clay content on the stabilization of soil organic matter (Parton et al., 1987).

Time

The time factor has received a great deal of attention in ecosystem ecology, perhaps in part because chronosequence studies of soils interact strongly with studies of primary succession (vegetation development on new substrates), which have a long history in ecology (cf., Clements, 1916). In fact, ecosystem theory is perhaps best developed for the understanding of changes in ecosystem function during primary succession and soil development (Vitousek & Walker, 1987); that theory in turn is driven primarily from analyses by soil scientists (cf., Walker & Syers, 1976).

One example of the use of chronosequences in ecological studies is in the taiga of Alaska, where Van Cleve et al. (1991) recently summarized the effects of the major state factors on forest ecosystem structure and function (see also, Van Cleve et al., 1986). As part of this overall study, they determined changes in ecosystem state and dynamics as a function of time in newly deposited river bars along the silt-laden Tanana River. Here, ecosystems develop from barely vegetated bars with a surface salt crust to tall forests. Several crucial transitions occur along the way (Van Cleve et al., 1991), including: (i) colonization of bare silt bars by plants and the deposition of plant litter on the soil surface, which causes transpiration to dominate over surface evaporation, thereby eliminating the surface salt crust and allowing further

ECOSYSTEM STRUCTURE & FUNCTION CONTROL FACTORS

colonization by plants; and (ii) accumulation of soil organic matter and the development of a carpet of feather moss on the forest floor late in the sequence insulate the soil sufficiently to allow the formation of permafrost, which in turn reduces soil-rooting volume, nutrient availability, and plant productivity.

Human Activity

Human activity was not separated as a major factor in *Factors of Soil Formation* (Jenny, 1941), although the book contains numerous examples of the effects of humans on soils and ecosystems. By 1980, the influence of human activity received a more explicit treatment (Jenny, 1980; see also Amundson & Jenny, 1991). In the 50 yr since the publication of *Factors of Soil Formation*, the human population has grown from <2 billion to 5.3 billion, fossil fuel consumption (as C) has increased from <2 Pg to 5.5 Pg/yr, and the use of N fertilizer has increased from <10 to about 80 Tg/yr (to give just a few examples). The impact of humanity on ecosystems has grown proportionately, and there can be no doubt that human activity is a major and increasing—but far from simple—state factor.

One example of how human activity has become an independent state factor is the alteration of atmospheric N inputs to terrestrial ecosystems. Substantial attention has been paid to N cycling and availability in ecosystem studies, because N supply often represents the proximate limitation to primary production in both natural and managed ecosystems (Vitousek & Howarth, 1991). However, as discussed above, N cycling generally is not an independent factor affecting ecosystem dynamics (outside of fertilized areas), but rather an internal process that is controlled (ultimately) by other state factors.

Human activity alters the global N cycle by fixing atmospheric N both deliberately and accidentally, and by mobilizing long-term storage pools of N from vegetation and soils into the hydrosphere and the atmosphere. The result of these alterations has been an enormous increase in the deposition of fixed N in precipitation in highly populated regions—to as much as 50 to 100 kg N ha/yr in some areas of Europe. Under these circumstances, N availability is no longer an internal process within ecosystems—it is an external, potentially independent cause of changes in ecosystem state. This enhanced anthropogenic deposition has been implicated in nutrient leaching, forest dieback, and a reduction of biological diversity in Europe (Schulze et al., 1989).

INTERACTIONS AND MODELING

As *Factors of Soil Formation* makes clear, it is only rarely that the influences of a single ecosystem state factor can be analyzed more or less in isolation in a climo-, bio-, topo-, litho-, or chronosequence. In most sites, several factors interact. Occasionally, the effects of the interactions of two or more state factors can be studied directly. For example, in 1991 Van Cleve et al.'s summary of taiga forest ecosystems (discussed above) considered the independent and interactive influence of several factors. Similarly, Vitousek et al. (1990, 1992) evaluated controls on physiological and ecosystem processes in an environmental matrix of lava flows on Mauna Loa Volcano, HI. The sites were such that the consequences of variation in climate (including independent gradients of temperature and moisture), parent material texture, time, and their interactions could be determined while holding organisms, relief, and parent material chemistry reasonably constant.

More commonly, the interactions of the major state factors are complex, and they interact with ecosystem states and processes at several levels of control from ultimate (interactions with other state factors) to proximate (cf., Robertson, 1988). In such situations, the interactions can only be addressed through modeling. Various modeling approaches have been employed, including a direct use of multiple regression based directly on the major state factors (Jenny, 1980). Perhaps the most promising current approach is the application of ecosystem-level simulation models, of which the CENTURY soil organic matter model (Parton et al., 1987, 1988) is a good example.

The CENTURY model originally was designed to simulate ecosystem dynamics of C, N, P, and S in North American Great Plains grasslands. The development of the model made explicit use of sequences of state factors—including the Great Plains temperature/precipitation matrix identified by Jenny (1941), variation in ecosystems along toposequences within the prairie, effects of soil texture on soil organic matter stabilization, the effects of agricultural land use on soil nutrient pools and fluxes, and the influence of time. However, the model does not (generally) use variations in ecosystem properties along gradients of state factors to calculate coefficients for a regression model. Rather, it uses such variation as a means towards understanding the processes that control ecosystem structure and function, and to design experiments where appropriate. The understanding so gained is then used to model those processes explicitly (where possible). The consequences of interactions of the factors can then be determined in a complex simulation model—with the real, complex ecosystems available for validation.

The CENTURY family of models is now being extended to other ecosystems, including tropical grasslands, savannas, and forests. In many cases, new controls and interactions must be built into the models—and therefore new gradients of the major state factors are being used to develop and to test the extended model (Parton et al., 1989).

APPLICATION TO GLOBAL CHANGE

One of the most significant challenges facing the scientific community is the understanding and ultimately the control of global environmental change. The approaches outlined by Jenny in *Factors of Soil Formation* can be—and in many cases already have been—useful to this effort.

One contribution of Jenny's approach is the more or less direct application of ecosystem state factors to understanding emissions of the radiatively or chemically active trace gases responsible for some components of global change (Matson et al., 1989). For example, patterns of N cycling in tropical forests are strongly affected by variation in temperature (= altitude), precipitation, soil fertility (primarily an interaciton of parent material and time), and of course human activity. Matson and Vitousek (1987) demonstrated that emissions of the radiatively active trace gas N_2O from tropical forest soil are strongly correlated with variations in N cycling along these gradients—they then established a sampling system to determine emissions in a range of sites. The ultimate result was a global source budget for intact tropical forests (Matson & Vitousek, 1990); it suggested that such forests produce 3 to 3.7 Tg of N_2O, the largest single source in the background global budget. The approach is now being applied to determining the effects of human-caused land-use change.

A second way that a state factor-based approach can be useful in studies of global change is in the development of ecosystem models (discussed above). Such models are fundamental to understanding and predicting change on coarse spatial or long temporal scales, and the use of state factors is crucial to their development and validation. For example, the CENTURY model includes routines that calculate fluxes of trace gases (Parton et al., 1988b); these are now being used to predict the consequences of changes in climate, the atmosphere, and land use (Schimel et al., 1990).

A third way that state factors can be useful is in the design of measurements and experiments to determine causes and consequences of global environmental change. The international scientific community is now developing an integrated research program, the International Geosphere-Biosphere Program (IGBP), to address global change. That program includes a project on Global Change in Terrestrial Ecosystems (GCTE) (including soils). The operational plan for GCTE includes a task designed to evaluate biogeochemical causes and consequences of global change (defined as climatic, atmospheric, and land-use change) in "critical regions." To quote the plan:

> Two of the three critical regions (tropical semi-arid systems and boreal forest-tundra transitions) are defined as transitions between ecosystems differing in their dominant life form.... These transitions result from biological modifications and feedbacks superimposed on a continual, gradual change in an environmental factor (precipitation in semi-arid tropical systems, temperature in boreal forest/tundra systems). Accordingly, the regulation of these transitions and their biogeochemistry can be examined most effectively by distributing measurements and experiments in sites along a gradient of the underlying environmental factor.... Where [comparative analyses] can be based on well-defined and continuous variation in an environmental factor—as in gradient studies—they can yield substantially greater insight into how that factor controls biogeochemical processes. Further, ecosystem-level experimentation that is replicated along an environmental gradient can be used to analyze interactions among the underlying environmental factor, other environmental variables, and biotic components of ecosystems. Finally, carryoug out research and modelling along gradients enforces an extensive, regional, and realistic perspective upon biogeochemical studies.

This plan clearly is based directly on the approach synthesized in *Factors of Soil Formation*, it illustrates that Jenny's insights will continue to be applied to ecosystems for decades to come.

REFERENCES

Aber, J.D., J.M. Melillo, K.J. Nadelhoffer, C.A. McClaugherty, and J. Pastor. 1985. Fine root turnover in forest ecosystems in relation to quantity and form of nitrogen availability: a comparison of two methods. Oecologia 66:317-321.

Amundson, R., and H. Jenny. 1991. The place of humans in the state factor theory of ecosystems and their soils. Soil Sci. 151:99-109.

Aplet, G.H. 1990. Alteration of earthworm community biomass by the alien *Myrica faya* in Hawaii. Oecologia 82:414-416.

Clements, F.E. 1916. Plant succession: An analysis of the development of vegetation. Publ. 242. Carnegie Inst., Washington, DC.

Grubb, P.J. 1977. Control of forest growth and distribution on wet tropical mountains. Annu. Rev. Ecol. Syst. 8:83-107.

Heaney, A., and j. Proctor. 1989. Chemical elements in litter in forests on Volcan Barva, Costa Rica. p. 255-271. In J. Proctor (ed.) Mineral nutrients in tropical forest and savanna ecosystems. Blackwell Sci., Oxford, England.

Jenny, H. 1941. Factors of soil formation. McGraw-Hill Book Co., New York.

Jenny, H. 1980. The soil resource: Origin and behavior. Springer-Verlag, New York.

Jenny, H. S.P. Gessel, and F.T. Bingham. 1949. Comparative study of decomposition rates of organic matter in temperate and tropical regions. Soil Sci. 68:419-432.

Körner, Ch. 1989. The nutrient status of plants from high elevation: a worldwide comparison. Oecologia 81:379-391.

Marrs, R.H., J. Proctor, A. Heaney, and M.D. Mountford. 1988. Changes in soil nitrogen-mineralization and nitrification along an altitudinal transect in tropical rainforest in Costa Rica. J. Ecol. 76:466-482.

Matson, P.A., and P.M. Vitousek. 1987. Cross-system comparisons of soil nitrogen transformations and nitrous oxide flux in tropical forest ecosystems. Global Biogeochem. Cycles 1:163-170.

Matson, P.A., and P.M. Vitousek. 1990. Ecosystem approach to a global nitrous oxide budget. BioScience 40:667-672.

Matson, P.A., P.M. Vitousek, and D.S. Schimel. 1989. Regional extrapolation of trace gas flux based on soils and ecosystems. p. 97-108. In M.O. Andreae and D.S. Schimel (ed.) Exchange of trace gases between terrestrial ecosystems and the atmosphere. John Wiley & Sons, Chichester, England.

Parton, W.J., A.R. Mosier, and D.S. Schimel. 1988b. Rates and pathways of nitrous oxide production in a shortgrass steppe. Biogeochemistry 6:45-58.

Parton, W.J., R.L. Sanford, P.A. Sanchez, and J.B. Stewart. 1989. Modeling soil organic matter dynamics in tropical soils. p. 153-171. In D.C. Coleman et al. (ed.) Dynamics of soil organic matter in tropical ecosystems. NifTAL Project, Univ. Hawaii, Honolulu.

Parton, W.J., D.S. Schimel, C.V. Cole, and D.S. Ojima. 1987. Analysis of factors controlling soil organic matter levels in Great Plains grasslands. Soil Sci. Soc. Am. J. 51:1173-1179.

Parton, W.J., J.W.B. Stewart, and C.V. Cole. 1988a. Dynamics of C, N, P, and S in grassland soils: A model. Biogeochemistry 5:109-131.

Pastor, J., and W.M. Post. 1986. Influence of climate, soil moisture, and succession on forest carbon and nitrogen cycles. Biogeochemistry 2:3-27.

Pastor, J., J.D. Aber, C.A. McClanghrety, and J.M. Melillo. 1982. Geology, soils and vegetation of Blackhawk Island, Wisconsin. Am. Midl. Nat. 108:266-277.

Pastor, J., J.D. Aber, C.A. McClaugherty, and J.M. Melillo. 1984. Aboveground production and N and P cycling along a nitrogen mineralization gradient on Blackhawk Island, Wisconsin. Ecology 65:256-268.

Peterjohn, W.T., and D.L. Correll. 1984. Nutrient dynamics in an agricultural watershed: Observations on the role of a riparian forest. Ecology 65:1466-1475.

Robertson, G.P. 1988. Nitrification and denitrification in humid tropical ecosystems: Potential controls on nitrogen retention. p. 55-69. In J. Proctor (ed.) Mineral nutrients in tropical forest and savanna ecosystems. Blackwell Scientific, Oxford, England.

Schimel, D.S., T.G.F. Kittel, A.K. Knapp, T.R. Seastedt, W.J. Parton, and V.B. Bryan. 1991. Physiological interactions along resource gradients in a tallgrass prairie. Ecology 72:672-784.

Schimel, D.S., W.J. Parton, C.V. Cole, D.S. Ojima, and T.G.F. Kittel. 1990. Grassland biogeochemistry: Links to atmospheric processes. Clim. Change 17:13-25.

Schimel, D.S., M.A. Stillwell, and R.G. Woodmansee. 1985. Biogeochemistry of C, N and P in a soil catena of the shortgrass steppe. Ecology 66:276-282.

Schulze, E.D., O.L. Lange, and R. Orem (ed.) 1989. Forest decline and air pollution: A study of spruce (*Picea abies*) on acid soils. Springer-Verlag, Berlin, Germany.

Sellers, P.J., F.G. Hall, G. Asrar, D.E. Strebel, and R.E. Murphy. 1988. The first ISLSCP field experiment (FIFE). Bull. Am. Meteorol. Soc. 69:22-27.

Van Cleve, K., F.S. Chapin III, P.W. Flanagan, L.A. Viereck, and C.T. Dyrness (ed.). 1986. Forest ecosystems in the Alaskan Taiga: A synthesis of structure and function. Springer-Verlag, New York.

Van Cleve, K., F.S. Chapin III, C.T. Dyrness, and L.A. Viereck. 1991. Element cycling in taiga forests: state-factor control. BioScience 41:78-88.

Vitousek, P.M., and R. Howarth. 1991. Nitrogen limitation on land and in the sea: How can it occur? Biogeochem. 13:87-115.

Vitousek, P.M., and L.R. Walker. 1987. Colonization, succession, and resource availability: Ecosystem-level interactions. p. 207-223. *In* A. Gray, M. Crawley, and P.J. Edwards (ed.) Colonization, succession, and stability. Blackwell Sci., Oxford, England.

Vitousek, P.M., and L.R. Walker. 1989. Biological invasion by *Myrica faya* in Hawaii: Plant demography, nitrogen fixation, and ecosystem effects. Ecol. Monogr. 59:247-265.

Vitousek, P.M., G. Aplet, D.R. Turner, and J.J. Lockwood. 1992. The Mauna Loa environmental matrix: Foliar and soil nutrients. Oecologia 89:372-382.

Vitousek, P.M., C.B. Field, and P.A. Matson. 1990. Variation in foliar $\delta^{13}C$ in Hawaiian *Metrosideros polymorpha*: A case of internal resistance? Oecologia 84:362-370.

Vitousek, P.M., L.R. Walker, L.D. Whiteaker, D. Mueller-Dombois, and P.A. Matson. 1987. Biological invasion by *Myrica faya* alters ecosystem development in Hawaii. Science (Washington, DC) 238:802-804.

Vogt, K., C.C. Grier, and D.J. Vogt. 1986. Production, turnover, and nutrient dynamics of above- and below-ground detritus of world forests. Adv. Ecol. Res. 15:303-377.

Walker, L.R., and P.M. Vitousek. 1991. Interactions of an alien and native tree during primary succession in Hawaii Volcanoes National Park. Ecology 72:1449-1455.

Walker, T.W., and J.K. Syers. 1976. The fate of phosphorus during pedogenesis. Geoderma 14:1-19.

Wedin, D.A., and D. Tilman. 1990. Species effects on nitrogen cycling: A test with perennial grasses. Oecologia 84:433-441.

Wessman, C.A., J.D. Aber, D.L. Peterson, and J.M. Melillo. 1988. Remote sensing of canopy chemistry and nitrogen cycling in temperate forest ecosystems. Nature (London) 335:154-156.

6 Soil Geography and Factor Functionality: Interacting Concepts

R. W. Arnold
Soil Survey Division
USDA-SCS
Washington, District of Columbia

ABSTRACT

The soil survey of the USA flourished under C.F. Marbut (1913–1934). His leadership transformed the ideas of Glinka, Hilgard, and others into a coherent U.S. pedological philosophy. There were two USDA soil inventory programs during the 1930s and 1940s—soil surveys under C.E. Kellogg and soil conservation surveys under E.A. Norton. The surveys were combined in 1951 and Dr. Kellogg continued to provide leadership from 1951 to 1974. According to Jenny (1941, p. 262) soil geographers were those scientists who developed maps, and soil functionalists were those scientists who developed curves and equations. Kellogg's philosophy for the soil survey embraced both geography and factor analysis. World War II delayed the impact of Jenny's book *Factors of Soil Formation* on U.S. pedological thought. In the decades since the 1940s field and laboratory investigations have repeatedly demonstrated the complexities and heterogeneity of soil patterns and soil genesis. Our understanding of the pedosphere benefited from the interaction of the methods of soil geography and those of soil factor functional analysis.

Tracing the development of ideas and their translation into working concepts in U.S. pedology is fascinating but uncertain because of varying experiences and interpretations of both the participants and the observers. What appears to be profound to one individual is commonplace to another individual. Important interactions and responses among individuals are seldom part of the literature; yet, our legacy relies on traditions that have been shared along the pathway of change.

Jenny pointed out that soil geographers assemble soil knowledge in the form of maps; whereas soil functionalists assemble soil knowledge in the form of curves or equations (Jenny, 1941, p. 262). These two methods of interpre-

Copyright © 1994 Soil Science Society of America, 677 S. Segoe Rd., Madison, WI 53711, USA. *Factors of Soil Formation: A Fiftieth Anniversary Retrospective*. SSSA Special Publication 33.

tation have been vital to the development and growth of the soil survey program in the USA.

Each of us grasps and tempers the facts of events and personalities thereby giving rise to the reality of our perceptions. Functional relationships between properties of soils and the complexities of their spatial environments continue to intrigue us and challenge our understanding of the pedosphere.

IN THE EARLY DAYS

The soil survey started under the leadership of Professor Milton Whitney in 1899 and used geological concepts of soils for some years. The geologic bias was still evident in Bureau of Soils Bulletin 85, *A Study of Soils of the U.S.* by Coffey (1912), and also in Bulletin 96, *Soils of the United States* by Marbut et al. (1913). Although soil was considered to be residual or transported and was subdivided by rock type or transport mechanism, Coffey (1912, p. 34) did suggest that soil was "...a natural body having a definite genesis and distinct nature of its own and occupying an independent position in the formations constituting the surface of the earth." The concept was there but not general acceptance.

In 1913, Dr. Curtis F. Marbut was appointed as soil scientist in charge of the soil survey. His training as a geologist, work with W.M. Davis, and his translation of Glinka's book on soil types and their development (Glinka, 1927) seem to have combined in such a way that pedology became a working discipline in the USA. By the time of the First International Congress of the Society of Soil Science, Marbut (1928) stated, "...it was found to be wholly impossible to harmonize fundamental soil characteristics with geologic characteristics on any other basis than the complete independence of the one from the other group of characteristics." He also pointed out that it was the American field soil scientist who defined and applied the ultimate soil unit which was based on the characteristics of the soil itself and not on the theories of formation.

In 1928 Marbut lectured at the U.S. Department of Agriculture Graduate School. Mimeographed copies of his transcribed lecture notes are still found in the libraries of a number of land-grant universities. Doctor Charles E. Kellogg, who took over the role of principal soil scientist of the Soil Survey Division, Bureau of Chemistry and Soils after Dr. Marbut's retirement in 1934, commented that the lectures were for him the most important of Marbut's written work, partly because they were put in his hands at the time of greatest need. He said the notes indicated how Marbut's mind worked, how he applied scientific correlation to soils, and the relative importance he gave to soil characteristics, as well as the various environmental factors working in combintion.

Several ideas that guided the early workers of the soil survey are still evident in the concepts that we use today. Marbut (1951, p. 25) said, "...the processes of soil development act on parent materials to make soil by continuing the decomposition of minerals, accumulating and assimilating organic matter, dissolving mineral and organic materials and leaching them from the soil, translocating materials within the soil thus developing soil horizons, forming new chemical compounds and by developing structure." Some years later Roy Simonson (1959) expanded these ideas and reformulated them into a generalized outline of soil genesis that has provided a meaningful framework for understanding the complex interactions of processes involved in soil development.

Doctor Wolfanger, a former soil surveyor, writing about Marbut's leadership in genesis, recalled that the idea "...given a type of soil forming material, the forces and factors operating on that material, rather than the material itself, are of prime importance in governing the resultant soil's fundamental properties," was a most revolutionary departure from the time-honored and widely held genetic interpretation. (Wolfanger, 1930, p. xi). He also wrote that the pedogenic cycle, whereby soils pass through cycles of development resembling those of other changing natural phenomena, was probably the most significant concept of modern soil science. Wolfanger (1930, p. 11) further noted that the application of this philosphy in the soil survey "...furnished a key to both the striking differences between certain soils which, on casual analysis, it would appear should have been alike...and close resemblances which should be dissimilar."

Geographic relationships are essential to mapping soils, thus Wolfanger (1930, p. xii) examined the major soil divisions described by the soil survey to set forth, as he said "...their fundamental functional characteristics, and to determine some of their essential and more important geographic relationships." The concept of a "mature" soil was related to the mature landscapes formed by downwearing, and the soils were characterized by (Wolfanger, 1930, p. 11) "...those features which all soils of a given environment acquire when comparatively undisturbed by erosion or deposition through an extended period of time on relatively level surfaces and under good drainage conditions."

The mature soil was one in which the conditions and forces of the environment had impressed themselves fully. It also was recognized that in different regions the mature soils, having developed to maturity under different kinds of environment have different characteristics. Marbut (1951, p. 28) explained that, "the relationship of the characteristics in different kinds of mature soils is not a relationship of stage of development, but a relationship of fundamental difference. They are different things produced by different processes and not different stages in the development of any one process."

Once when asked about the major influence of natural vegetation on soils, Marbut discussed the prairie soils in the Corn Belt region where there are similar conditions except for the natural vegetation. His reasoning (Marbut, 1951, p. 21) for "...the justification of drawing such a conclusion of course is based on the assumption that when we have eliminated the geological, topographic and climatic factors we have eliminated all of the important soil making influences except natural vegetation." But in the tradition of a good scientist, he (Marbut, 1951, p. 22) went on to say, "...while so far as we know up to the present time the forces eliminated plus the influence of natural vegetation constitute all of the important soil making forces, yet it is perfectly possible that in the future some other soil making forces not included in these may be discovered which will demonstrate to us that the natural vegetation in the prairies has had much less to do with the determination of the characteristics of its soils than we now think." Additional applications of this experimental method approach in pedology were described by Jenny (1941) as an appropriate technique to study the influence of each of the soil-forming factors.

Shortly after Dr. Kellogg was designated as the principal soil scientist in charge of the soil survey division in the Bureau of Chemistry and Soils, his investigations on the *Development and Significance of the Great Soil Groups of the United States* (Kellogg, 1936) were published. It was his intent to summarize recent knowledge about the formation and importance of major soil groups for the general reader. He integrated the concepts of zonal and intrazonal soils with a schematic outline of the pedogenic cycle, and provided a functional soil-forming factor equation. Doctor Kellogg (1936, p. 9) mentioned that "...as a graphic illustration of those factors of the landscape most prominently associated with and responsible for the development of soil the following can be written: Soil = f(climate, vegetation, relief, age, parent rock)." Obviously, Kellogg and Jenny had been teaching and writing about many of the same concepts.

THE DUAL SOIL SURVEYS

A Soil Erosion Service was created in the Department of Interior in 1933 in response to the disturbing billows of dust streaming all the way to the Atlantic ocean. In 1935, the activities were expanded, the name changed to the Soil Conservation Service (SCS) and responsibilities transferred to the Department of Agriculture (Simonson, 1987). Because soil conservation was considered in its broadest sense to imply permanent maintenance of the productive capacity of the land, it was necessary that "...land be used for purposes for which it was best suited and also necessitated the adoption of such soil conservation practices as required for each kind of land (Norton, 1939, p. 1)." The conservation plans were to be developed in accordance with a physical inventory, particularly of soil conditions, percentage of slope, character and degree of erosion, and present land use. Thus, the SCS established a Physical Surveys Division to carry out soil conservation surveys.

A joint Committee on Soil and Erosion Surveys was established in 1937 to coordinate requests for soil surveys thereby avoiding duplication and ensuring that the information obtained would meet the requirements for basic information needed in the programs of the department. When the Soil Survey Division of the Bureau of Chemistry and Soils published the *Soil Survey Manual* in 1937 (Kellogg, 1937) this document guided the scientific work of both of the soil surveys in the Department of Agriculture.

Although Dr. Kellogg acknowledged the assistance of Ableiter, Baldwin, Carter, Hearn, Lapham, McKericher, Nikiforoff and Rice in preparing the *Soil Survey Manual*, it is obvious that much of the philosophy was his. The manual noted that it was essential for nomenclature and terminology to be as uniform as possible and that some standardization of methods was needed in order that the work could be correctly interpreted everywhere (Kellogg, 1937, p. 2). It also pointed out that the methodology needed to be sufficiently flexible so that all types of soils and landscapes could be mapped with scientific accuracy, practical effectiveness, and expediency, and that new ideas for improvement could be developed.

The manual stressed that although the fundamental purpose of the work was to serve practical objectives, it had to be based on sound scientific principles. Doctor Kellogg (1937, p. 6) emphasized that "...the purpose and function of the soil survey was to map the soils and other physical features of the land in a manner such that the problems of land use planning could be solved rationally in the best interests of the individual or social groups responsible for their solution." The practicality of purpose and the value of a scientific foundation are very much a part of the mission of the National Cooperative Soil Survey in the 1990s (Soil Survey Staff, 1992a).

With regard to other uses of the soil survey, Dr. Kellogg (1937, p. 10) said, "it is perhaps unnecessary that attention be drawn specifically to the great contributions of the work to soil science, as scientific investigations are a necessary prerequisite to the construction of the soil map itself."

The concept of the "mature" soil had been changed to the "normal" soil—one having a profile in equilibrium with the two principal forces of the environment—native vegetation and climate, with the other factors—relief, parent material, and age in a neutral position. Doctor Kellogg (1937, p. 66) reiterated that the normal soil may not be dominant in an area and in some areas there may not be any true normal soil, for example, where the entire area is low and poorly drained. It was suggested that the field men establish the principal properties of the normal soil by examining profiles at as many points as possible and preferably in areas of different parent materials. Once the normal soil was identified, the field investigator was to define the local units that differed significantly from the normal soil as a result of differences in one or more of the five factors of genesis (Kellogg, 1937, p. 67). Here was a description of how to use functional relationships to help locate, describe, and define important soils and soil variability in a landscape. The geographic method of understanding soils relied on defining each soil unit in terms of mappable differentiating internal and external features. These features served as a basis for determining the areal extent of the unit and for drawing its boundaries. Empirical correlations of observable internal soil features and external landscape features could be made, even if genetic causes and effects were unknown.

During the late 1930s, the *Soil Survey Manual* not only guided the conduct of the soil survey, it also provided an important philosophical basis for many of its activities. Consider this (Kellogg, 1937, p. 68), "...the soil description must be entirely objective. Any suggestions regarding genesis or any other speculative explanation of the facts should be kept strictly apart from the description of these facts. The description of soil profiles are the fundamental data of soil science." The soil survey also classified soils and it was Dr. Kellogg's belief (1937, p. 68) that "...the classification of soils is predicated on a knowledge of their morphology, and unless that morphology is precisely known, classification is impossible."

In summary, the concept of soil encompassed the total expression of all the forces comprising its environment, and field investigations were to include the geographic setting and special features of the landscape. The pieces were in place—soil geography was moving ahead rapidly; however, quantification and critical evaluation of functional analysis were not major activities of the soil survey program.

SHARING KNOWLEDGE

Pedology was becoming a discipline with formal courses of instruction mainly in agronomy and geography departments. Familiar textbooks in 1940 were *Pedology* by J.S. Joffe (1936) at Rutgers University and *The Nature and Property of Soils*, 3rd Ed. by T.L. Lyon and H.O. Buckman (1938) at Cornell University; however, many professors developed their own lecture notes drawing on concepts from Dr. Kellogg, Soviet articles presented at the 1st International Congress of Soil Science, and the 1938 *Yearbook of Agriculture on Soils* (USDA, 1938). Such personal lecture notes permitted the professors to more fully express concepts and relationships based on their experiences.

For many young scientists who had associations with soil survey activities such as Marlin Cline, Francis Hole, Frank Riecken, Eugene Whiteside, and Eric Winters, the *Soil Survey Manual* of 1937 was not only a standard reference, it also was a book of important concepts.

When Professor Jenny's book *Factors of Soil Formation* became available for classroom use in late 1941, no one imagined that soon the USA would devote its major energies to war. There was little time to assess the state-of-the-art of pedology. However, by the late 1940s new attitudes and wartime experiences set in motion ideas that would significantly affect the soil survey.

The World Soil Geography unit under the initial leadership of Professor Cline demonstrated the utility of examining interactions of the soil factors to predict kinds of soils, their properties, and expected behavior. Jenny's formalized factorial equation (Jenny, 1941, p. 17) fixed indelibly the universality of "functional relationships" thereby helping to consolidate the existing concepts and models common to most U.S. pedologists.

It is evident that Professor Jenny was up-to-date with the literature of the times. His book on the *Properties of Colloids* published by the Stanford University Press in 1938, was soon followed by *Factors of Soil Formation: A System of Quantitative Pedology* (Jenny, 1938, 1941). Jenny eloquently made the bridge from the site factors of soils to those of the environment noting that soil climate (cl'), soil organisms (o'), and soil relief (r') were each functionally related to those same factors in the environment; and therefore, the fundamental equation of soil forming factors became $s = f(cl, o,$

INTERACTION OF SOIL GEOGRAPHY AND FACTOR FUNCTIONALITY

$r, p, t...$) whereby soil characteristics were functionally related to the factors of the environment (Jenny, 1941, p. 16, 17). Because of the difficulty of recognizing parent material, Jenny (1941, p. 53) suggested that parent material might be better estimated as the initial state of the soil system rather than referring to the strata below the soil which might or might not be similar to the original parent material.

He said that to be meaningful, the equation of soil formation needed to be solved, meaning that the indeterminate function "f" must be replaced by some specific quantitative relationship. The total change of any soil property depended on all the changes of all the soil forming factors.

In his book, Jenny assembled soil data into a comprehensive scheme based on numerical relationships. He believed (Jenny, 1941, p. 262) that soil properties were correlated with independent variables commonly called "soil forming factors" and that such an approach would assist in the understanding of soil differentiation as well as help explain the geographical distribution of soil types.

The ultimate goal of functional analysis, according to Jenny (1941, p. xi), was the formulation of quantitative laws that permit mathematical treatment. In a prophetic statement he said, "...as yet, no correlation between soil properties and conditioning factors has been found under field conditions which satisfies the requirements of generality and rigidity of natural laws. For that reason, the less presumptuous name, 'functional relationship' was chosen" (Jenny, 1941, p. xii). There still are no natural laws in pedology.

To describe the connection with soil geography, Jenny said (1941, p. 262) "the goal of the soil geographer is the assemblage of soil knowledge in the form of a map. In contrast, the goal of the functionalist is the assemblage of soil knowledge in the form of a curve or an equation." Jenny noted that soil maps display areal arrangement but give no insight into causal relationships and that curves reveal dependency of soil properties on soil-forming factors but conversion of such knowledge to field conditions is impossible without the soil map. He said, (1941, p. 262) "clearly it is the union of the geographic and the functional methods that provides the most effective means of pedological research." In other words, the spatial patterns of the soil survey need to be understood through functional relationships with the soil-forming factors in space and time.

THE COMMON GROUND

In retrospect, it is evident that the dynamic activities and ideas of pedology were merging and gaining strength in the late 1930s and the early 1940s (Simonson, 1948, Part 5, p. 19–22). There was general agreement on the five major soil-forming factors and that their interactions produced the processes that differentiated parent materials into soil properties and horizons. There was general agreement that different soils could form from the same parent materials and that similar soils could form from different kinds of parent materials. There was general agreement that soil properties and characteristics could be empirically related to segments of the landscapes in which they occur.

Soil genesis was thought of as the set of explanations about how soils developed and that genesis involved more than observing facts in the field and the laboratory. It also was generally believed that additional scientific investigations would elucidate the cause-and-effect relationships which functionally connect soil properties to environmental factors and conditions, both past and present.

Some very interesting pedologic studies were taking place at the time Jenny's textbook became available. Simonson (1991) reported on buried soils formed from till in Iowa reinforcing the concern about polygenetic soils expressed by Jenny (1941, p. 79–81). Guy Smith completed his M.S. with Jenny at Missouri (Jenny & Smith, 1935) and then undertook his classic study with R.S. Smith at Illinois providing a pedologic interpretation of the properties and distribution of loess in Illinois (Smith, 1942).

The ideas were all there, in bits and pieces, and being used by those who would guide the soil survey in the decades ahead. Professor Jenny brought together so-called "hard evidence" clearly demonstrating that functional relationships could be quantified and that such results would help us better understand the linkages of soils to the factors of soil formation and the importance of their interactions.

Reviews and predictions were part of the golden anniversary of the soil survey in 1949. Doctor Kellogg (1950, p. 9) reconfirmed a basic principle, namely, that "theoretically each combination of the five genetic factors—climate, living matter, parent material, relief, and time—produces the same profile." He also stated that with careful work it was possible to go a long way toward explaining the processes that give rise to contrasting soils from similar materials. Doctor Kellogg (1950, p. 9) said that soil scientists have a responsibility for discovering genetic laws and relationships because we didn't understand our soils "... until we know how they formed and why one varies from another." He also pointed out that as the soil survey expands its activities it must provide for fundamental research, using both the experimental method and the method of scientific correlation, otherwise it might promise more than it could deliver.

Considerably later, it was suggested that soils following different pathways of development may have the same or similar properties at some point in time, and that the use of multiple working hypotheses could be a useful technique to guide genetic interpretations (Arnold, 1965).

A matrix approach to functional relationships provided biotoposequences as hypotheses. Field discovery of new series in the matrix supported Jenny's concept of soil as a continuum (Riecken, 1965). Many soil survey legends of the Midwest attest to the power of using biotoposequences to explain certain landscapes and to account for some mapping inclusions.

In my opinion, a major contribution of Professor Jenny to the soil survey was his presentation of basic concepts of soil formation in a clear and concise way that enabled other teachers and researchers to portray pedology as a viable dynamic subject amenable to systematic research and quantification. I dare say that most of us who are more than 50 yr old were products of such academic training, we all talk and write about the fundamental soil-forming factor equation. Those of us who have taught continued the tradition, and we have a similarity of thought processes and experiences with soils that I find truly remarkable. Yes, the time was right and Dr. Hans Jenny was on the spot, filled a niche, and helped make it possible for many of us to see beyond the first course in soil science. We are grateful for those opportunities that have let soil science and especially pedology advance to its present status.

ALONG THE WAY

At the Silver Anniversary of the Soil Science Society of America in 1960, Professor Cline (1961) described important changes in the model of soil noting that three things, in his opinion, shaped many of the changes. These involved improved understanding and application of the concepts of geomorphology, time as a factor in soil genesis, and processes of soil formation. The linkage to, and similarity with, some of Professor Jenny's ideas are, by now, common knowledge.

The Soil Survey Division of the Bureau of Plant Industry, Soils and Agricultural Engineering was combined with that of the Soil Conservation Service in late 1951 under the leadership of Dr. Kellogg (Simonson, 1987, p. 23). The Investigations Staff under the leadership of Guy Smith, undertook four major geomorphology–soil landscape research projects in the 1950s. They were the Greenfield Quadrangle in Iowa studying loess-till statigraphy and paleosols, the Coastal Plain project in North Carolina, the Willamette Valley project in Oregon, and the Desert project in New Mexico. These studies demonstrated the universality of pedimentation and its importance in recognizing and delineating the complexity of soil parent materials. Interruption of geomorphic cycles with attendant landscape erosion and deposition of pedisediments was observed to be common in most landscapes.

As more of the USA was mapped and pedological research increased it became apparent that time was a complicated factor in soil formation. The implications of the Pleistocene Epoch on climate, sea level, sedimentation, and landscape evolution altered the thoughts about the nature of normal soils. In some areas more than one normal soil was thought to occur, eventually they were recognized as early stages of development rather than expressing the full impact of climate and vegetation (Cline, 1961). Radiocarbon dating and detailed profile studies brought out time relationships that did not always fit the earlier concepts of how and when soils had formed.

The older terminology and concepts of soil processes were streamlined with the integrated interactions that were so succinctly outlined by Simonson (1959). Disparities in thinking about how processes proceed were brought together enabling us to be more aware that soils are the products of processes interacting with materials and if we interpret soils correctly there are very few anomalies.

Doctor Cline brought into keen perspective the role that the model of soil had in making soil surveys and pedology more quantitative (Cline, 1977). For example, *Soil Taxonomy* (Soil Survey Staff, 1975) was conceived in 1950, put into practice in 1965, and published in 1975 (Soil Survey Staff, 1975). In 1992, the fifth edition of updated *Keys to Soil Taxonomy* was published (Soil Survey Staff, 1992) to help keep the world current with recent changes in the system. The continual refinement of definitions and diagnostics are hallmarks of the quantification of soil classification.

CONCLUSION

It is interesting that the newer versions of the *Soil Survey Manual*, (Soil Survey Staff, 1951, 1993) the *National Soils Handbook*, (Soil Survey Staff, 1983) and *Soil Taxonomy* (Soil Survey Staff, 1975) are all authored by the soil survey staff instead of designated individuals. The major influence in pedology and in the soil survey today is that of collective judgments and purposeful guidance. Documentation, quantitative databases, and critical evaluation are important aspects of current National Cooperative Soil Survey activities.

As I view the soil survey some five decades after Professor Jenny's book on the *Factors of Soil Formation* I am confident that the union of the geographic and the functional methods have been providing an exceptionally effective means of pedological research. There have been meaningful interactions of the geographical research of the soil survey and the soil factor relationships of the functionalists. Together they have made progress. They have opened our eyes and our minds to far more complexity than was anticipated. The men and women of the National Cooperative Soil Survey are world class pedologists whose understanding of spatial patterns and geographic pedogenesis is a notable achievement in the history of soil science. It has been an exciting 50 yr.

REFERENCES

Arnold, R.W. 1965. Multiple working hypothesis in soil genesis. Soil Sci. Soc. Am. Proc. 25:717–724.

Cline, M.G. 1961. The changing model of soil. Soil Sci. Soc. Am. Proc. 25:442–446.

Cline, M.G. 1977. Historical highlights in soil genesis, morphology, and classification. Soil Sci. Soc. Am. J. 41:250–254.

Coffey, G.N. 1912. A study of soils of the United States. USDA Bur. of Soils Bull. 85. U.S. Gov. Print. Office, Washington,D C.

Glinka, K.D. 1927. The great soil groups of the world and their development. Translation by C.F. Marbut. Edwards Brothers, Ann Arbor, MI.

Jenny, H. 1938. Properties of colloids. Stanford Univ. Press, Stanford Univ., Palo Alto, CA.

Jenny, H. 1941. Factors of soil formation. McGraw Hill Book Co., New York.

Jenny, H., and G.D. Smith. 1935. Colloid chemical aspects of claypan formation in soil profiles. Soil Sci. 39:377–389.

Joffe, J.S. 1936. Pedology. Rutgers Univ. Press, New Brunswick, NJ.

Kellogg, C.E. 1936. Development and significance of the great soil groups of the United States. USDA Misc. Publ. 229. U.S. Gov. Print. Office, Washington, DC.

Kellogg, C.E. 1937. Soil survey manual. USDA Misc. Publ. 274. U.S. Gov. Print. Office, Washington, DC.
Kellogg, C.E. 1950. The future of the soil survey. Soil Sci. Soc. Am. Proc. 14:8–13.
Lyon, T.L., and H.O. Buckman. 1938. The nature and property of soils. 3rd ed. Macmillan, New York.
Marbut, C.F. 1928. A scheme for soil classification. p. 1–31. *In* R.E. Deemer et al. (ed.) Trans. Comm. 5, 1, Int. Congress Soil Sci. Am. Organization Comm., Washington, DC.
Marbut, C.F. 1951. Soils: Their genesis and classification. SSSA, Madison, WI.
Marbut, C.F., H.H. Bennett, J.E. Lapham, and M.H. Lapham. 1913. Soils of the United States. U.S. Bur. of Soils Bull. USDA Soils Bull. 96. U.S. Gov. Print. Office, Washington, DC.
Norton, E.A. 1939. Soil conservation survey handbook. USDA Misc. Publ. 352. U.S. Gov. Print. Office, Washington, DC.
Riecken, F.F. 1965. Present soil forming factors and processes in temperate regions. Soil Sci. 99:58–64.
Simonson, R.W. 1941. Studies of buried soils formed from till in Iowa. Soil Sci. Soc. Am. Proc. 6:373–381.
Simonson, R.W. 1959. Outline of a generalized theory of soil genesis. Soil Sci. Soc. Am. Proc. 23:152–156.
Simonson, R.W. 1987. Historical aspects of soil survey and soil classification. Soil Surv. Horiz. 27:1–10.
Smith, G.D. 1942. Illinois loess-variations in its properties and distribution: a pedologic interpretation. Illinois Agric. Exp. Stn. Bull. 490.
Soil Survey Staff. 1951. Soil survey manual. USDA Agric. Handb. 18. U.S. Gov. Print. Office, Washington, DC.
Soil Survey Staff. 1975. Soil taxonomy: A basic system of soil classification for making and interpreting soil surveys. USDA-SCS Agric. Handb. 436. U.S. Gov. Print. Office, Washington, DC.
Soil Survey Staff. 1983. National soil survey handbook. USDA-SCS Agric. Handb. 430-VI-NSH. U.S. Gov. Print. Office, Washington, DC.
Soil Survey Staff. 1992a. Program plan, soil survey division. USDA-SCS U.S. Gov. Print. Office, Washington, DC.
Soil Survey Staff. 1992b. Keys to soil taxonomy. 5th ed. SMSS Tech. Monogr. 19. Pocahontas Press, Inc., Blacksburg, VA.
Soil Survey Staff. 1993. Soil survey manual. USDA Agric. Handb. U.S. Gov. Print. Office, Washington, DC.
U.S. Department of Agriculture. 1938. Soils, yearbook of agriculture. U.S. Gov. Print. Office, Washington, DC.
Wolfanger, L.A. 1930. The major soil divisions of the United States. John Wiley & Sons, New York.

7 Soil Formation Theory: A Summary of Its Principal Impacts on Geography, Geomorphology, Soil-Geomorphology, Quaternary Geology and Paleopedology

D. L. Johnson

University of Illinois
Urbana, Illinois

Francis D. Hole

University of Wisconsin
Madison, Wisconsin

ABSTRACT

Soil formation theory and philosophy was born in eastern Europe during the late nineteenth century as the intellectual offspring of V.V. Dokuchaev and his students and colleagues. In the early twentieth century it diffused into western Europe and North America where it became fundamental theory for such practical applications as the USDA soil mapping program, and various intellectual efforts, such as Hans Jenny's eloquent and substantive treatise *The Factors of Soil Formation*. The framework also provided the philosophical foundation for such fundamental—albeit historically problematic—concepts as normal and zonal soils, monogenetic and polygenetic soils, and paleosols. It formed a major part of the philosophy behind chronosequence and paleopedological studies, and the concept of polygenetic landscapes, and was an important part of the philosophy behind climatic geomorphology as well as aspects of process geomorphology. In fact, the formational paradigm has been omnicient as the soil-theoretical foundation of the earth sciences during this century. A negative entry exists in this otherwise positive ledger in that the framework lacks visibility for two important theoretical and factual aspects of pedogenesis—soil evolution theory, and biomechanical processes. In these collective ways the framework has significantly impacted the disciplines of geography, geomorphology, soil geomorphology, Quaternary geology, and paleopedology, which includes undergraduate teaching and graduate training in these fields. Hans Jenny's role and image as the

---

Copyright © 1994 Soil Science Society of America, 677 S. Segoe Rd., Madison, WI 53711, USA. *Factors of Soil Formation: A Fiftieth Anniversary Retrospective.* SSSA Special Publication 33.

principal and eloquent advocate of the formational-factorial paradigm, among other accomplishments, has won him an honored and deserved place in twentieth century science.

In this paper, we summarize some principal impacts, both pros and cons, of soil formational-factorial theory on geography, geomorphology, soil-geomorphology, Quaternary geology, and paleopedology. We offer one caveat. Because the literature and intellectual linkages between soil genesis and the fields indicated are so vast, subtle, and intricate, it is presumptuous that we could possibly identify and comprehend them all. Nevertheless, we have tried to the best of our judgement and experience. But, because a detailed treatment of these linkages would doubtless be of book length, we attempt only to summarize them here. Satisfaction of detail is, thus, lost to the reader, and for that we apologize.

Part of our strategy is to identify those papers and treatises which have significantly impacted these fields because they incorporated soil formation theory as showcased in H. Jenny's landmark book (Jenny, 1941) and in a coeval paper by J. Thorp (1941). We begin by examining the intellectual ambience that existed at the time Jenny's book and Thorp's paper appeared. We then briefly review and summarize the impacts of formational theory on the five fields indicated—these impacts may be considered the main pros, or more or less positive impacts of the framework. We then discuss the main cons of formational theory. We end with a summary and conclusions of the principal impacts of formation theory.

INTELLECTUAL AMBIENCE IN SOILS PRIOR TO 1941

It is noteworthy that the intellectual ambiance that prevailed in 1941 was probably optimal for a positive reaction to Jenny's book and it's message. The reason was that by 1941 the five factors approach had already become *the* principle paradigm of soil genesis. Indeed, the five factors approach had appeared in a number of key monographs, papers, and textbooks on soils dating from the 19th century, first in Russian, later in German and English (e.g., Dokuchaev, 1893, 1898, 1899; Ramann, 1911; Glinka, 1914, 1927; Neustruev, 1927; Marbut, 1928 and 1935, unpublished data; Joffe, 1936; Kellogg, 1934, 1936; Byers et al., 1938; Thorp, 1941; and references within). In addition to emphasizing the formational-factorial approach, this body of literature, which culminated in the 1930s and early 1940s, represented the essential pedogenetic theory of the time. This largely Russian approach was endorsed and widely promulgated not only by academic giants in the field, such as Joffe whose popular book *Pedology* was published in 1936, but also by respected personnel of one of the most influential—in terms of power, resources, and visibility—governmental organizations, the USDA. In par-

SOIL FORMATION THEORY

ticular, the contributions of Marbut (1935), Kellogg (1936), Byers et al. (1938), and Thorp (1941) espoused the factorial-formational approach by authority of the USDA.

The intellectual and theoretical "seedbed" for Jenny's book and Thorp's paper had, thus, long been prepared by academic and governmental authority and served to ensure their success. No competing theory then existed, and the remarkably bold idea of Dokuchaev and his colleagues in reducing the incredible complexities of soil genesis to only five factors—four exogenous environmental factors, plus time—had and still has wide appeal. The formational-factorial model was generally adopted by the international soil science community and became essential theory in pedology and in most soil mapping programs. The fact that endogenous soil properties and conditions evolve in soils which often affect or control their subsequent genetic pathways more or less independently of the four exogenous environmental factors, that is, that soils *evolve*, was acknowledged by Dockuchaev and his followers (Kossovich, 1911, Nikiforoff, 1942, 1949; Neustruev, 1927; Rode, 1947), but apparently not by USDA personnel, nor by Jenny. As will be shown, this was one of two key genetic omissions from the formational-factorial framework. The second key omission is that the "*O*" factor in the formational framework was traditionally equated almost exclusively with plants, or more specifically the biochemical role of plants in soil genesis. The biomechanical role of plants, animals, and other lifeforms was omitted from the formational framework even though an abundant literature on the subject existed during its formative period.

Though Thorp (1941) produced a major paper on the factorial-functional framework the same year Jenny's book appeared, Jenny is the one that history most closely associates with the model. This is because Jenny's distinctive contribution was to theoretically and methodologically *showcase* the formational-factorial approach by: using clear and simple language, using many excellent illustrations, bringing together under one cover numerous examples of soil forming situations, and calling attention to potential quantitative applications in pedology. In fact, Jenny performed for quantitative pedology what Horton (1945) and Strahler (1950) later did for quantitative geomorphology. Clearly, Jenny's book was an auspicious and constructive entry into the 1940s, a decade otherwise marked by the inauspicious and destructive aspects of war.

IMPACTS OF SOIL FORMATION THEORY ON GEOGRAPHY

The impact of the five factors model on geography has been greatest in the teaching of physical geography and soil geography at both the introductory and advanced levels. For example, we randomly picked 10 introductory physical geography texts off our shelves. In each we found that the soil geography sections were either partly or dominantly structured around soil formational-factorial theory. The texts are: *Elements of Geography* (Trewartha et al., 1957), *Physical Geography* (Strahler, 1969), *Physical Geography*

(Patton et al., 1974), *Introduction to Physical Geography* (Gabler et al., 1975), *Contemporary Physical Geography* (Navarra, 1981), *Essentials of Physical Geography Today* (Oberlander & Muller, 1982), *Physical Geography: Earth Systems and Human Interactions* (Miller, 1985), *Physical Geography: A Landscape Appreciation* (McKnight, 1987), *Essentials of Physical Geography* (Scott, 1991), and *Physical Geography: An Introduction to Earth Environments* (Bradshaw & Weaver, 1993). Further, several texts from our shelves that focus either entirely on soil geography, or contain a significant component of it, have incorporated the five factors approach. These include: *The Geography of Soil* (Bunting, 1965), *World Soils* (Bridges, 1970), *A Geography of Soils* (Basile, 1971), *The Geography of Soils* (Steila, 1976), *Principles and Applications of Soil Geography* (Bridges & Davidson, 1982), and *Soil Genesis and Classification* (Buol et al., 1973, 1980, 1989).

As espoused in these introductory and advanced texts, soil formation theory obviously has had a long and important effect on the training of undergraduates and graduate professionals in physical and soil geography, and in other fields as well. We have long been involved in teaching introductory and advanced courses on these subjects wherein soil genesis is packaged, at least partly, in soil formation theory. In fact from the instructor's standpoint (and the student's), what could be simpler or easier at the introductory or advanced level than presenting the incredible complexities of soil genesis in terms of five easily comprehensible elements of our everyday world. The formational-factorial framework epitomizes the point that the simpler the model, other things being equal, usually the more attractive the model is to users (Johnson & Watson-Stegner, 1987).

IMPACTS OF SOIL FORMATION THEORY ON GEOMORPHOLOGY, SOIL-GEOMORPHOLOGY, AND QUATERNARY GEOLOGY

The Contributions of K. Bryan and C.C. Albritton

In 1943, 2 yr after Jenny's book and Thorp's paper appeared, K. Bryan, a geomorphologist, and C. Albritton, an archaeologist, produced a seminal theoretical paper in pedology titled "Soil Phenomena as Evidence for Climatic Change" (Bryan & Albritton, 1943). This paper drew heavily and collectively on soil formation-factorial theory as espoused by Jenny and Thorp. On the basis of citation frequency, Bryan and Albritton's (1943) conceptual world view of pedogenesis and geomorphogenesis seems to have been as much influenced by Thorp's paper as Jenny's book. It also drew on the concepts of "mature" and "normal" soils as espoused by USDA personnel under the leadership of Marbut (1928, unpublished data; 1935) and Kellogg (1936, 1937), and the largely Russian-inspired concepts of normality, zonality, paleopedology, and soil evolution (Dokuchaev, 1893; Polynov, 1927, p. 1–33; Neustruev, 1927). Marbut (1928, unpublished data), apparently influenced by the views of W.M. Davis (Bryan's former professor), proposed for soil classification purposes, the concept of soil "maturity" and the cyclical nature of soils, developing from youth to senility (Olson, 1989).

In addition to other aspects of soil genetic theory, Bryan and Albritton (1943) formulated such concepts as monogenetic vs. polygenetic soils, composite soils, and precocious soils. Their paper had a tremendous impact on pedology and geomorphology. The concept of polygenesis in particular has had a dramatic and lasting affect on these fields. That is because polygenesis, though conceived in a soils framework, became conceptually fused with geomorphology in a paper by Peltier (1950) titled "The Geographical Cycle in Periglacial Regions as Related to Climatic Geomorphology." Peltier's paper, which drew heavily on Bryan and Albritton's concept of polygenesis, was well received, and polygenesis soon became the theoretical core of the geomorphic concept of "polygenetic landscapes." This framework gave impetus to the fledgling subfield of climatic geomorphology which burst into maturity in the late 1950s as expressed in the landmark paper by Büdel (1957). A synonym of polygenesis, "polycyclicity" later appeared in European literature (Duchaufour, 1982). Without soil formation theory, as showcased by Thorp's paper and Jenny's book in 1941, and without the polygenetic conceptualizations of Bryan and Albritton in 1943, these developments probably either would not have occurred in geomorphology, or would have occurred differently.

Bryan and Albritton's concept of monogenetic soils has, likewise, left a profound mark in chronosequence work. Jenny (1941) reasoned that if a site is selected where parent materials are the same but are on surfaces of different age, as in a flight of river terraces, and where the factors of climate, vegetation, and relief have not changed, the variation in soils will be a function of time. For this to be valid, each older soil in the chronosequence would have had to have followed identical similar pedogenic pathways indicated by the properties of the younger soils (Vreeken, 1975). In other words, the soils would have to be "monogenetic" in origin, that is, climate (but also vegetation and relief) must have remained constant during soil development. Apart from the validity or invalidity of this line of reasoning (i.e., the counterpoint that all soils are polygenetic), the concept led to a profusion of chronosequence studies and the formulation of quantitative indices to support them (Birkeland, 1992, with references).

To summarize, one theoretical pathway went from soil formation theory as showcased by Jenny and Thorp in 1941, to the concept of polygenetic soils as formulated by Bryan and Albritton in 1943, thence to the concept of polygenetic landscapes advanced by Peltier in 1950, which served to invigorate the evolving field of climatic geomorphology in the early-mid 1950's (e.g., Büdel, 1957, 1982). Another pathway led from Thorp's paper and Jenny's book, to Bryan and Albritton's concept of monogenetic soils, and thence to soil chronosequence studies, a line of research that has proliferated in geomorphology and Quaternary geology in recent decades.

The Contributions of P.W. Birkeland

A geomorphic contribution of significance that incorporates the soil formation paradigm is the volume by Birkeland (1974), titled *Pedology, Weathering and Geomorphological Research*, and its update *Soils and Geomorphology* (1984). Jenny's approach forms the conceptual and methodological core of Birkeland's book, which is deemed a geologist's explanation of the five factors of soil formation. That the book has been used by many students as a virtual handbook in soil-geomorphology and Quaternary geology is testimony to the present influence of formation theory in these fields. The historical circumstances here are no accident. Before writing his book, Birkeland was a professional colleague of Jenny's in the Department of Soils and Plant Nutrition at the University of California at Berkeley in the 1960s. At one point, Birkeland took over Jenny's pedology course when the latter was on leave, an experience that left indelible pedogenetic impressions on him (Birkeland, personal communication, 1991). One only has to consult the text and References sections of almost any recent book or monograph on geomorphology, soil-geomorphology, Quaternary geology, or even introductory general earth sciences to appreciate the impact of Birkeland's books—and by implication, soil formation theory—on these fields (e.g., Birkeland & Larson, 1989; Catt, 1986; Gerrard, 1981; Knuepfer & McFadden, 1990; Martini & Chesworth, 1992; Tarbuck & Lutgens,1976).

Where the five factors framework, Jenny's 1941 book, and the concept of monogenesis provided the theoretical inspiration for chronosequence studies, Birkeland and his students provided the practical examples. A consequence was a plethora of chronosequence studies that followed the appearance of Birkeland's book in 1974 (Birkeland, 1984, 1992).

IMPACTS OF SOIL FORMATION THEORY ON PALEOPEDOLOGY

Impacts on Terminology, Definitions and Theory

The impact of soil formation theory on the rapidly growing subfield of paleopedology has been, in our opinion, profound. It is reflected partly in the early usage and or definitions of the terms "fossil soil," "relict soil," "paleopedology," "monogenesis," and "polygenesis." These terms and their definitions involve the most fundamental theoretical concepts of paleopedology, and each—either in early or later usage, has at least some conceptual linkage to soil formation theory. A brief historical sketch of these terms and their use and/or definitions will demonstrate the linkages. The following exposition, part of which is modified from Johnson (1993b), also demonstrates a history of term misuse and abuse in paleopedology.

The term "fossil soil" was until about 1950 more or less synonymous with buried soil (Ramann, 1911, 1928, p. 36; Krokos, 1923; Polynov, 1927, p. 33; Nikoforoff, 1943; Geze, 1947; Williams, 1945; Frye, 1949; Thorp, 1949;

Thorp et al., 1951). One exception we noted was by Thorp (1935) who believed that some fossil soils may remain unburied (i.e., some surfaces were fossil soils). However, in later writings, at least through 1951, he abandoned this view and equated fossil soils to buried soils (e.g., Thorp et al., 1951).

The term "relict soil" ("reliktenboden") was defined by Ramann in 1911 in *Bodenkunde* (p. 526) for "soils which have lost their original properties under the influence of a change in climate" (see also Ramann, 1928, p. 37). Polynov (1927, p. 1–33), apparently unaware of this definition, defined two categories of paleopedological phenomena (i.e., two kinds of soils) in a paper titled "Contributions of Russian Scientists to Paleopedology" that are the approximate equivalents to relict soil as defined by Ramann. He called these kinds of soils "secondary soil formations" and "two-stage soils," both of which could be produced by climatic change, as well as other changes of environment. Thorp et al. (1951), also apparently unaware of Ramann's definition and Polynov's redefinition, redefined relict soils as "the products of former environmental conditions, but modified by the soil-forming factors of today" [note the influence of formational theory in all three statements, in Ramann's case three decades before Jenny's book appeared]. Thorp et al. (1951) also used the terms relict soil and polygenetic soil interchangeably, as synonyms.

First use of the term "paleopedology" apparently was by Polynov (1927, p. 1–33) who included it in the title (cited above) of a seminal paper on the subject. Though poorly written in English, this difficult to read paper contains several concepts of formational theory which significantly influenced subsequent pedological, paleopedological, and geomorphological thought. For example, it contains the seeds of the concepts of monogenesis (i.e., "normal soils") and polygenesis (i.e., soils affected by climatic changes) which were later formalized by Bryan and Albritton (1943), concepts which impacted both the field of climatic geomorphology, as we have seen, and definitions of "paleosol" in the 1950s. Polynov included both surface and buried soils within the domain of paleopedology.

In 1943, Nikiforoff produced an essay titled "Introduction to Paleopedology" that, like Polynov's paper, also was rooted in formational theory. Also like Polynov, Nikiforoff's conceptual domain of paleopedology included buried soils plus those surface soils that are "out of harmony with the environment."

Interestingly, while Polynov's and Nikiforoff's essays were on paleopedology, neither used the word "paleosol." First use of the term "paleosol" appears to have been by Erhart in 1932 who used it in reference to buried loessal soils in the Alsace of France (Erhart, 1932). He later used the term to describe a soil buried by basalt in the Bas-Languedoc of southern France (Erhart, 1940). In both cases Erhart used the term "paleosol" to describe *buried* soils. Thorp (1949) citing Erhart (1940) introduced "paleosol" to the English language with the same meaning. Thus up until the late 1940s the terms "fossil soil," "buried soil," and "paleosol" were essentially synonymous and nonproblematic.

A transition in terms and meanings, however, occurred between 1947 and 1965 that would create conceptual confusion thereafter. First, drawing on formational theory, and perhaps inspired by Nikiforoff's and Polynov's broadly perceived domain of paleopedology and Bryan and Albritton's concept of polygenesis—all also rooted in formational theory, Gèze (1947) expanded "paleosol" to include not only buried soils but also surface soils that contained "expressions of ancient soil formation" (sols de formation ancienne). Following on this expansion, Hunt and Sokoloff (1950) used the term for surface soils with morphologies that reflect "ancient rather than modern environments." By these usages, Gèze and Hunt and Sokoloff described forms of soils similar to Ramann's (1911) "relict soil" definition, similar to Polynov's "secondary" and "two-stage" soils, similar to Bryan and Albritton's polygenetic soil, and similar to Thorp et al.'s redefinition of relict soil. Then, 6 yr later Ruhe, oblivious to early usage, redefined a paleosol as a buried soil that had formed on a landscape during the geologic past, which owing to burial became preserved as a "fossil" soil (Ruhe, 1956). He also said that if such a buried soil became exhumed due to erosive stripping, it should still be designated a paleosol because it would now be foreign to its present environment [note the influence of formation theory in both the redefinitions of Gèze, of Hunt and Sokoloff, and of Ruhe]. Two years later Ruhe and Daniels (1958) called this kind of once-buried soil an *exhumed* paleosol. Eight years later Ruhe (1965), apparently unaware of Ramann's original definition of relict soil, Polynov's "secondary" and "two-stage" soils, nor Thorp et al.'s (1951) redefinition, further redefined *"relict soils"* as paleosols "that formed on pre-existing landscapes that were not buried." Relict soils thus became unburied paleosols.

Owing to these various redefinitions where priority of usage was almost totally ignored, conceptual confusion was created that exists to the present. Subsequent workers perpetuated the confusion by not only ignoring the original definitions but by variously misquoting and misparaphrasing the redefinitions as well (e.g., Valentine & Dalrymple, 1976; Wright, 1986; Yaalon, 1971). Other workers added to the confusion by further redefining paleosols to mean "soils of obvious antiquity" (Morrison, 1967, p. 10), "ancient soils" (Butzer, 1971, p. 170), "soils formed on a landscape of the past" (Valentine & Dalrymple, 1976; Yaalon, 1971; p. 29), "soils formed under an environment of the past" (Yaalon, 1983), "soils formed before the last cold period of the Pleistocene" (Duchaufour, 1982), and so on. Later workers also variously redefined relict soils to mean those soils "showing two or more sets of properties...related to different combinations of soil-forming factors through sets of often incompatible soil-forming processes [that] contain features formed during two or more periods with different environmental conditions" (Bronger & Catt, 1989a), "those that are anomalous in the present, soil-forming environment" (Tarnocai & Valentine, 1989), and so on. Aside from confusion created by this terminological and conceptual mess, the impact of formation theory is reflected in language chosen for all these redefinitions.

Other Impacts

Other impacts of formation theory on paleopedology are clearly demonstrated by reference to several recent books on the subject. For example, *Soils of the Past* by Retallack (1990), is divided into three parts, one of which focuses exclusively on the five-factors paradigm (6 chapters out of 21 total, or about 30% of the book). *Paleopedology*, edited by Bronger and Catt (1989b), is loaded with references to Jenny's book, to monogenetic and polygenetic soils, to relict soils, and to other aspects of formation theory. A debt to formation theory also is apparent in statements by the authors of several entries in the recent book *Weathering, Soils & Paleosols* edited by Martini and Chesworth (1992).

THE PROS AND CONS OF SOIL FORMATION THEORY

The Pros

The main pros of soil formation theory are the impacts on the theoretical and methdological developments of physical geography, geomorphology, soil–geomorphology, Quaternary geology, and paleopedology as outlined above. This assumes that, except for the problem of confused paleopedological terms noted, they were all positive impacts. However, since the pros as outlined above were not weighed and coevaluated with the cons (as outlined below) in this historical development, we lack a scale for comparing positive vs. negative impacts. We can only speculate what theoretical and methodological developments and differences might have occurred, and what fundamental truths arrived at, had the formational model been more flexible, or had viable competing theory existed and effectively packaged in 1941. What is obvious is that later competing theory—where process and systems linkages were emphasized over factors (Simonson, 1959, Hole, 1961, Runge,1973), encouraged a re-examination of pedogenesis, which led to a significantly broadened and strengthened theoretical base for pedology (e.g., Chesworth, 1973a,b; Huggett, 1975; Paton, 1978; Stegner, 1980, unpublished data; Smeck et al., 1983; Johnson & Watson-Stegner, 1987; Johnson et al., 1990).

The Cons

It was earlier noted that the main cons of the soil formation paradigm are the failure to capture several essential elements of theory. One such element is the notion (or fact) that soil morphological properties and conditions (i.e., pedogenic accessions) can evolve such that they themselves by their very presence, will cause profiles to change more or less independently of the exogenous environmental factors. This point of theory is not new, having long been either emphasized or alluded to by others (e.g., Cline, 1961; Kossovich, 1911; Kovda et al., 1968; Nikiforoff, 1942, 1949, 1955, 1965; Novak et al., 1971; Ruellan, 1971; p. 7-8; Samoylova, 1971; Yaalon, 1971, p. 36). Marbut (1935, p. 16) called attention to pedogenic accessions in his contribution to the *Atlas of American Agriculture*—both he and Nikiforoff (1942) referred to them as "acquired characteristics." An example might be a subsoil horizon that gradually develops such that it increasingly acts as an aquitard or aquiclude to downward moving water, ultimately resulting in water moving laterally within the soil as subsurface throughflow. The concept that soil properties and evolved accessions can effect minor and major changes in pedogenic pathways without changes in climate or vegetation has no visibility in soil formation theory, nor was the latter theory formulated to accommodate such concepts. That soil properties, and thus the soil, can change without changes in the "soil forming factors" invalidates the concept of monogenesis and polygenesis as originally formulated by Bryan and Albritton (1943). (A redefinition of polygenesis is given in Johnson et al., 1990.)

It is noteworthy that in the 1920s when the formational-factorial theory of the Russian school was absorbed by USDA personnel and infused in their evolving framework to aid in soil survey work, soil evolution theory was omitted. As indicated, the latter was an important part of early twentieth century Russian pedogenetic thought (see discussions in Nikiforoff, 1942, 1949; Rode, 1947). Had the USDA adopted it, would the ensuing historical development of pedology, soil-geomorphology, and paleopedology in North America and elsewhere have been different? Would Jenny's message in his 1941 treatise have been different?

A second genetic element that limits formational theory is an absence of conceptual visibility for biomechanical processes in the *"O"* factor. The biotic processes of pedogenesis, the *"O"* factor in soil formation theory, are both *biochemical* and *biomechanical* in nature. Whereas biochemical processes—mainly those related to plants, have been acknowledged for centuries as important in soil genesis, for example as showcased in Jenny's book, biomechanical processes have not. They were first emphasized by Darwin, who showcased them in *his* landmark book on pedology, titled *The Origin of Vegetable Mould through the Action of Worms* (Darwin, 1881). The book spawned numerous observations of biomechanical processes in soil genesis by scholars in various disciplines for several decades after 1881 (as cited in Johnson, 1993a).

However, the pedogenic significance of biomechanical processes owed to animals, plants, and other lifeforms was short-lived inasmuch as such processes failed to find visibility in soil formation theory then being formulated in Russia by Dokuchaev and his followers. This omission was carried into Europe, via Glinka's (1914) treatise, and into North America and the USDA via Marbut's (1927) translation of it, and ultimately into the Jenny and Thorp factorial worldview of soil genesis as expressed in their respective 1941 contributions. The omission is archtypified in Jenny's chapter on organisms, which gives useful visibility to biochemical processes only. No mention is made of the biomechanical roles of biota as documented earlier by such prominent figures as Branner (1896, 1900), Darwin (1881), de Chételat

(1938), Drummond (1885, 1888), Grinnell (1923), Hilgard (1906), Holmes (1893), Joffe (1936), Keller (1887), Lutz (1940), Lutz and Griswold (1938), Passarge (1904, p. 289-303), Seton (1904), Shaler (1888, 1891), Troll (1936, p. 275-312), and von Ihering (1882), nor is their work on the subject cited by either Jenny or Thorp. In fact, the "O" factor in soil formational-factorial theory, the very one that defines Earth surface processes as fundamentlaly different from those on all other planets and celestial bodies, is user-limiting because it connotes only plant-soil biochemistry. Not only are biomechanical processes not given useful visibility, they in general—with one exception we noted (Joffe, 1936, p. 103-115), have been totally ignored in soil formation theory.

We submit that the main reason for the tradition of equating plant biochemistry with the "O" factor is because pedology, which supposedly defines the scientific study of soils, has *never* functioned as a true science, but historically has been prostrated to the practical realities of agriculture. This fact, long realized and lamented by some (e.g., Marbut, 1921; Nikiforoff, 1959), has many subtle manifestations, one of which is that biomechanical processes, while very important *scientifically* (i.e., pedogenetically), appear insignificant *agriculturally* relative to biochemical processes. Here it is worth noting that agricultural-biochemical emphases permeated the Soil Survey Division of the USDA during the 1920 to 1940 formative period of soil formation theory in the USA. It also is notable that Jenny, in the first instance, was an agricultural chemist and Thorp was long associated with the USDA. Finally, it is worth reflecting on the fact that generations of geographers, geomorphologists, soil-geomorphologists, Quaternary geologists and paleopedologists, not to mention pedologists and soil scientists, have been trained with these mindsets.

SUMMARY AND CONCLUSIONS

Formation theory was born in eastern Europe in the latter part of the nineteenth century under Dokuchaev's influence and promulgation. It diffused to western Europe in the early twentieth century via treatises by Ramann (1911, 1918) and Glinka (1914), and to North America in the 1920s via Marbut's English translation of Glinka's treatise. The pedogenetic role of climate and its effect on vegetation was emphasized over all other factors in this volume. Under Marbut's leadership as Director of the Soil Survey Division of the USDA, soil formation theory—sans Russian views of soil evolution theory, came to form the theoretical base for soil survey work in the USA. Concepts of "mature" and "normal" soil evolved in the 1930s, and were advanced by USDA personnel following the leadership of Marbut and Kellogg. In 1941 two respected scientists, Thorp of the USDA and Jenny the academician, published major statements on formation theory. Jenny's book showcased the theory and attempted to establish it on a quantitative footing. Soil formation theory early in this century carried the seeds of "relict soil," "fossil soil," "paleopedology," "monogenesis," 'polygenesis," and soil chronosequence theory.

In 1927, Polynov wrote a key conceptual paper on paleopedology that largely reflects formational theory. Drawing on the views of Polynov and Jenny and Thorp, Bryan and Albritton formalized concepts of monogenetic and polygenetic soils in 1943. Such views were augmented in a paleopedological paper by Nikiforoff that same year. The concept of polygenesis, framed in formation theory, transferred to geomorphology through Peltier's paper in 1950, and indirectly to paleopedology during the period 1947 to 1965 via papers by Gèze, Hunt and Sokoloff, and by Ruhe in which redefinitions of "paleosol" without acknowledgements of priorities, were given. All these terms were linked by language and conceptuality to soil formation theory. These various concepts led to significant impacts in the undergraduate teaching, graduate research training, and thought linkages of students in physical geography, soil geography, geomorphology, soil-geomorphology, Quaternary geology, and paleopedology.

The main pros of soil formation theory, apart from confused paleopedological terms (which partly reflects the failure of practitioners to properly review the literature), are its impacts as summarized above. Its legacy is profound, especially in the fields of Quaternary geology, paleopedology, and in the teaching of physical geography. The main cons are that it carries no visibility for two key elements of theory, namely soil evolution theory, and the role of biomechanical processes in the *"O"* factor.

Soil formation theory has dominated, almost exclusively, the soil-theoretical base of the earth sciences during most of this century. Indeed, it has been omniscient in the pedogenetic underpinnings of not only the five fields indicated, but also of soil science, pedology and archaeology (e.g., see other papers, this volume). Hans Jenny's role and image as an eloquent advocate of this framework, in addition to his many other accomplishments, has won him a permanent and deserved place in twentieth century science.

It is, however, in the nature of science for established frameworks to be challenged by competing theory. But for competing theory to succeed it must explain phenomena either better, or differently in a useful way—or better *and* differently in useful ways. Formation theory *is* being challenged by competing theory, and we are authors to part of the challenge. Will competing theory replace formation theory? Probably not, but it will undoubtedly strengthen, diversify and improve the soil-theoretical base of the earth sciences. With this, we are certain, Hans Jenny would surely have agreed.

ACKNOWLEDGEMENTS

We thank Diana Johnson for critically evaluating this paper, Barbara Bonnell for typing it; and Ron Amundson, Jennifer Harden, and Mike Singer for improving it through their perceptive criticisms and editing.

REFERENCES

Basile, R.M. 1971. A geography of soils. W. C. Brown, Dubuque, IA.

Birkeland, P.W. 1974. Pedology, weathering, and geomorphological research. Oxford Univ. Press, London and New York.

Birkeland, P.W. 1984. Soils and geomorphology. Oxford Univ. Press, New York.

Birkeland,P.W. 1992. Quaternary soil chronosequences in various environments—extremely arid to humid tropical. p. 261-281. *In* I.P. Martini and W. Chesworth (ed.) Weathering, soils & paleosols. Elsevier, Amsterdam.

Birkeland, P.W. and E.E. Larson. 1989. Putnam's geology. Oxford Univ. Press, New York.

Branner, J.C. 1896. Decomposition of rocks in Brazil. Geol. Soc. Am. Bull. 7:255-314.

Branner, J.C. 1900. Ants as geological agents in the tropics. J. Geol. 8:151-153.

Bradshaw, M., and R. Weaver. 1993. Physical geography: An introduction to earth environments. Mosby, St. Louis, MO.

Bridges, E.M. 1970. World soils. Cambridge Univ. Press, Cambridge, England.

Bridges, E.M., and D.A. Davidson (ed.). 1982. Principles and applications of soil geography. Longman, New York.

Bronger, A. and J.A. Catt. 1989a. Paleosols: Problems of definition, recognition and interpretation. p. 1-7. *In* A. Bronger and J.A. Catt (ed.) Paleopedology: Nature and application of paleosols. Catena Suppl. 16. Catena Verlag, Cremlingen-Destedt, Germany.

Bronger, A., and J.A. Catt (ed.). 1989b. Paleopedology: Nature and application of paleosols. Catena Suppl. 16. Catena Verlag, Cremlingen-Destedt, Germany.

Bryan, K., and C.C. Albritton. 1943. Soil phenomena as evidence for climate changes. Am. J. Sci. 241:469-490.

Büdel, J. 1957. The double plantation surfaces in the humid tropics (In German). Z. Geomorphologie 2:201-228.

Büdel, J. 1982. Climatic geomorphology. (Translated from German by L. Fischer and D. Busche). Princeton Univ. Press, Princeton, NJ.

Bunting, B.T. 1965. The geography of soils. Aldine Publ. Co., Chicago.

Buol, S.W., F.D. Hole, and R.J. McCracken. 1973. Soil genesis and classification. Iowa State Univ. Press, Ames, IA.

Buol, S.W., F.D. Hole, and R.J. McCracken. 1980. Soil genesis and classification. 2nd ed. Iowa State Univ. Press, Ames, IA.

Buol, S.W., F.D. Hole, and R.J. McCracken. 1989. Soil genesis and classification. 3rd ed. Iowa State Univ. Press, Ames, IA.

Butzer, K.W. 1971. Environment and archaeology. 2nd ed. Aldine, Chicago.

Byers, H.G., C.E. Kellogg, M.S. Anderson, and J. Thorp. 1938. Formation of soil. p. 948-978. *In* Soils and men: Yearbook of agriculture. U.S. Gov. Print. Office, Washington, DC.

Catt, J.A. 1986. Soils and quaternary geology: A handbook for field scientists. Clorendon Press, Oxford, England.

Chesworth, W. 1973a. The parent rock effect in the genesis of soil. Geoderma 10:215-225.

Chesworth, W. 1973b. The residua system of chemical weathering: A model for the chemical breakdown of silicate rocks at the surface of the earth. J. Soil Sci. 24:69-81.

Cline, M.G. 1961. The changing model of soil. Soil Sci. Soc. Am. Proc. 25:442-446.

Darwin, C. 1881. The formation of vegetable mould through the action of worms. Appleton, New York.

de Chételat, E. 1938. Le modelé latéritique de l'ouest de la Guineé Française. Rev. Géog. Phys. Géol. Dynam. 11:5-120.

Dokuchaev, V.V. 1893. The Russian steppes: The study of soil in Russia, its past and present. Publ. for World's Columbian Exposition, Chicago, 1893. Dep. of Agric. Ministry of Crown Domains, St. Petersburg, Russia.

Dokuchaev, V.V. 1898. The problem of the re-evaluation of the land in European and Asiatic Russia. [In Russian.] Moscow.

Dokuchaev, V.V. 1899. On the theory of natural zones. [In Russian.] St. Petersburg, Russia.

Drummond, H. 1885. On the termite as the tropical analogue of the earth-worm. Proc. R. Soc. (Edinburgh) 13:137-146.

Drummond, H. 1888. Tropical Africa. Hodder and Houghton, London.

Duchaufour, P. 1982. Pedology. George Allen and Unwin, London. (Translated from French by T.R. Paton.)

Erhart, H. 1932. On the nature and genesis of paleosols in ancient loess of the Alsace. (In French.) Acad. Sci. Compt. Rend. 194:554–556.

Erhart, H. 1940. On the occurrence of a Quaternary paleosol in the Bas-Languedoc and the volcanic soils of the region. (In French.) Acad. Sci. Compt. Rend. 211:401–403.

Frye, J.C. 1949. Use of fossil soils in Kansas Pleistocene stratigraphy. Trans. Kansas Acad. Sci. 52:478–482.

Gabler, R.E., R. Sager, S. Brazier, and J. Pourciau. 1975. Introduction to physical geography. Rinehart Press, San Francisco, CA.

Gerrard, A.J. 1981. Soils and landforms: An integration of geomorphology and pedology. George Allen and Unwin, London.

Gèze, B. 1947. Paleosols and modern soils. p. 210–219. (Translated from French by D.L. Johnson.) In Compt. Rend. Conf. Pedol. Medit., Alger-Montpellier. Paris.

Glinka, K.D. 1914. Die typen der bodenbildung. Gebruder Borntrager, Berlin.

Glinka, K.D. 1927. The great soil groups of the world and their development (Translated by C.F. Marbut.) Edwards Brothers, Ann Arbor, MI.

Grinnell, J. 1923. The burrowing rodents of California as agents in soil formation. J. Mammal. 4—137-149.

Hilgard, E.W. 1906. Soils. The Macmillan Co., London.

Hole, F.D. 1961. A classification of pedoturbations and some other processes and factors of soil formation in relation to isotropism and anisotropism. Soil Sci. 91:375–377.

Holmes, W.H. 1893. Vestiges of early man in Minnesota. Am. Geol. 1:219–240.

Horton, R.E. 1945. Erosional development of streams and their drainage basins; Hydrophysical approach to quantitative morphology. Bull. Geol. Soc. Am. 56:275–370.

Huggett, R.J. 1975. Soil landscape systems: A model of soil genesis. Geoderma 13:1–22.

Hunt, C.B. and V.P. Sokoloff. 1950. Pre-Wisconsin soil in the Rocky Mountain Region, a progress report. U.S. Geol. Surv. Prof. Pap. 221-G. p. 109–123. U.S. Gov. Print. Office, Washington, DC.

Jenny, H. 1941. Factors of soil formation. McGraw Hill Co., New York.

Joffe, J.S. 1936. Pedology. Rutgers Univ. Press, New Brunswick, NJ.

Johnson, D.L. 1993a. Biomechanical processes and the Gaia paradigm in a unified pedogeomorphic and pedoarchaeologic framework: Dynamic denudation. In J.E. Foss et al. (ed.) Proc. Int. Conf. on Pedo-archaeology, 1st. Orlando, FL. February 1992. Univ. Tennessee Agric. Exp. Stn., Knoxville, TN.

Johnson, D.L. 1993b. The 'essential lexicon' of paleopedology: Problematic definitions and use, misuse, and abuse of key terms. In L.R. Follmer and D.L. Johnson (ed.) Paleopedology: Scope and domain. Proc. Int. Symp. Paleopedology, 2nd. 8 to 12 August. Champaign-Urbana, IL.

Johnson, D.L. and D. Watson-Stegner. 1987. Evolution model of pedogenesis. Soil Sci. 143:349–366.

Johnson, D.L., D. Watson-Stegner, D.N. Johnson, and R.J. Schaetzl. 1987. Proisotropic and proanisotropic processes of pedoturbation. Soil Sci. 143:278–292.

Johnson, D.L., E.A. Keller, and T.K. Rockwell. 1990. Dynamic pedogenesis: New views on some key soil concepts, and a model for interpreting Quaternary soils. Quat. Res. 33:306–319.

Keller, C. 1887. Reisebilder aus ostafrika und Madagaskar. G.F. Winter, Leipzig, Germany.

Kellogg, C. E. 1934. The place of soil in the biological complex. Science (Washington, DC) 39:46–51.

Kellogg, C.E. 1936. Development and significance of the great soil groups of the United States. USDA Misc. Publ. 229. U.S. Gov. Print. Office, Washington, DC.

Kellogg, C.E. 1937. Soil survey manual. USDA Misc. Publ. 274. U.S. Gov. Print. Office, Washington, DC.

Knuepfer, P.L.K., and L.D. McFadden (ed.). 1990. Soils and landscape evolution. Elsevier, Amsterdam.

Kossovich, P.A. 1911. Principles of soil science. Part II. Sect. 1. (In Russian.) St. Petersburg.

Kovda, V.A., B.G. Rozanov, and E.M. Samoylova. 1968. Soil map of the world. (In Russian.) Priroda 12:2–7.

Krokos, V. 1923. Loess and fossil soils of south-western Ukraina. Bull. Agric. Sci. Parts 3 to 4.

Lutz, H.J. 1940. Disturbance of forest soil resulting from the uprooting of trees. Yale Univ. School For. 45:1-37.
Lutz, H.J. and F.S. Griswold. 1938. The influence of tree roots on soil morphology. Am. J. Sci. 237:389-400.
Marbut, C.F. 1921. The contribution of soil surveys to soil science. Proc. Soc. Promotion Agric. Sci. 41:116-142.
Martini, I.P., and W. Chesworth. (ed.). 1992. Weathering, soils & paleosols. Elsevier, Amsterdam.
McKnight, T.L. 1987. Physical geography: A landscape appreciation. 2nd ed. Prentice-Hall, Englewood Cliffs, NJ.
Miller, E.W. 1985. Physical geography. Earth systems and human interactions. Charles E. Merrill Publ. Co., London.
Morrison, R.B. 1967. Principles of Quaternary soil stratigraphy. p. 1-69. *In* R.B. Morrison and H.E. Wright, Jr. (ed.) Quaternary soils. Proc. Int. Assoc. Quat. Res. Congress, 7th. vol. 9. Center for Water Resour.; Desert Res. Inst., Univ. Nevada, Reno.
Navarra, J.B. 1981. Contemporary physical geography. Saunders College Publ., New York.
Neustruev, S.S. 1927. Genesis of soils. p. 1-98. *In* Russian pedological investigations. 3rd ed. Acad. Sci., Leningrad.
Nikiforoff, C.C. 1942. Soil dynamics. Am. Sci. 30:36-50.
Nikiforoff, C.C. 1943. Introduction to paleopedology. Am. J. Sci. 241:194-200.
Nikiforoff, C.C. 1949. Weathering and soil evolution. Soil Sci., 68:219-230.
Nikiforoff, C.C. 1955. Hardpan soils of the coastal plain of southern Maryland. Geol. Surv. Prof. Pap. 267-B. U.S. Gov. Print. Office, Wash., DC.
Nikiforoff, C.C. 1959. Reappraisal of the soil. Science (Washington, DC) 129:186-196.
Nikiforoff, C.C. 1965. Pedogenic criteria of climatic changes. p. 189-200. *In* H. Shapley (ed.) Climatic change: Evidence, causes, and effects. Harvard Univ. Press, Cambridge, England.
Novak, R.J., H.L. Motto, and L.A. Douglas. 1971. The effect of time and particle size on mineral alteration in several Quaternary soils in New Jersey and Pennsylvania, U.S.A. p. 241-224. *In* D.H. Yaalon (ed.) Paleopedology: Origin, nature and dating of paleosols. Kefer Press, Jerusalem, Israel.
Oberlander, T.M., and R.A. Muller. 1982. Essentials of physical geography today. Random House, New York.
Olson, C.G. 1989. Soil geomorphic research and the importance of paleosol stratigraphy to Quaternary investigations, midwestern USA. p. 129-142. *In* A. Bronger & J.A. Catt (ed.) Paleopedology: Nature and application of paleosols. Catena Suppl. 16. Catena Verlag, Cremlingen-Destedt, Germany.
Passarge, S. 1904. The Kalahari. (In German.) Dietrich Reimer, Berlin.
Paton, T.R. 1978. The formation of soil material. George Allen and Unwin, London.
Patton, C.P., C.S. Alexander, and F.L. Kramer. 1974. Physical geography. 2nd ed. Duxbury Press, North Scituate, MA.
Peltier, L.C. 1950. The geographic cycle in periglacial regions as it is related to climatic geomorphology. Ann. Assoc. Am. Geogr. 40:214-236.
Polynov, V.V. 1927. Contributions of Russian scientists to paleopedology. Russian Pedol. Invest. Part 8. Acad. Sci. Moscow.
Ramann, E. 1911. Bodenkunde. 3rd ed. J. Springer, Berlin.
Ramann, E. 1918. Bodenbildung und Bodeneinteilung. J. Springer, Berlin, Germany.
Ramann, E. 1928. Evolution and classification of soils. (Translated from German by C.L. Whittles.) W. Heffer and Sons, Ltd. Cambridge, England.
Retallack, G. 1990. Soils of the past: An introduction to paleopedology. Unwin Hyman, Inc., Boston.
Rode, A.A. 1947. The soil forming process and soil evolution. (Translated from Russian by J.S. Joffe). Israel Program of Sci. Transl., Jerusalem.
Ruellan, A. 1971. The history of soils: Some problems of definition and interpretation. p. 3-27. *In* D.H. Yaalon (ed.) Paleopedology: Origin, nature and dating of paleosols. Keter Press, Jerusalem.
Ruhe, R.V. 1956. Geomorphic surfaces and the nature of soils. Soil Sci. 82:441-455.
Ruhe, R.V. 1965. Quaternary paleopedology. p. 755-764. *In* H.E. Wright and D.G. Frey (ed.) Quaternary of the United States. Princeton Univ. Press, Princeton, NJ.

Ruhe, R.V., and R.B. Daniels. 1958. Soils, paleosols, and soil-horizon nomenclature. Soil Sci. Soc. Am. Proc. 22:66–69.

Runge, E.C.A. 1973. Soil development sequences and energy models. Soil Sci. 115:183–193.

Samoylova, E.M. 1971. Some relict signs in contemporary soils of the Tambov Lowland, U.S.S.R. p. 173–179. *In* D.H. Yaalon (ed.) Paleopedology. Keter Press, Jerusalem.

Scott, R.C. 1991. Essentials of physical geography. West Publ. Co., St. Paul, NY.

Seton, E.T. 1904. The master plowman of the west. Century Mag. 68:300–307.

Shaler, N.S. 1888. Animal agency in soil-making. Pop. Sci. Monthly 32:484–487.

Shaler, N.S. 1891. The origin and nature of soils. p. 213–345. U.S. Geol. Surv. 12th Ann. Rep. 1890–1891, Part 1. U.S. Geol. Surv., Washington, DC.

Simonson, R.W. 1959. Outline of a generalized theory of soil genesis. Soil Sci. Soc. Am. Proc. 23:152–156.

Smeck, N.E., E.C.A. Runge, and E.E. Mackintosh. 1983. Dynamics and genetic modelling of soil systems. p. 51–81. *In* L.P. Wilding et al. (ed.) Pedogenesis and soil taxonomy I: Concepts and interactions. Elsevier Sci., Amsterdam.

Steila, D. 1976. The geography of soils. Prentice Hall, Englewood Cliffs, NJ.

Strahler, A.N. 1950. Equilibrium theory of erosional slopes approached by frequency distribution analysis. Am. J. Sci. 248:673–696, 800–814.

Strahler, A.N. 1969. Physical geography. 3rd ed. John Wiley & Sons, New York.

Tarbuck, E.J. and F.K. Lutgens. 1976. Earth sciences. Charles E. Merrill Publ. Co., Columbus, OH.

Tarnocai, C., and K.W.G. Valentine. 1989. Relict soil properties of the Arctic and Subarctic regions. p. 9–39. *In* A. Bronger and J.A. Catt (ed.) Paleopedology: Nature and application of paleosols. Catena Suppl. 16. Catena Verlag, Cremlingen-Destedt, Germany.

Thorp, J. 1935. Soil profile studies as an aid to understanding recent geology. Bull. Geol. Soc. China 14:360–393.

Thorp, J. 1941. The influence of environment on soil formation. Soil Sci. Soc. Am. Proc. 6:39–46.

Thorp, J. 1949. Interrelations of Pleistocene geology and soil science. Geol. Soc. Am. Bull. 60:1517–1526.

Thorp, J., W.M. Johnson, and E.C. Reed. 1951. Some post-Pleistocene buried soils of central United States. J. Soil Sci. 2:1–19.

Trewartha, G.T., A.N. Robinson, and E.H. Hammond. 1957. Elements of geography. Mcgraw-Hill Book Co., New York.

Troll, C. 1936. Termite savannas: Studies on vegetation and landscape science of the tropics. Part 2. (In German.) Landerkundliche Forschung Festschrift Norbert Krebs. J. Engelhorns Nachf, Stuttgart, Germany.

Valentine, K.W.G., and J.B. Dalrymple. 1976. Quaternary buried paleosols; A critical appraisal. Quat. Res. 6:209–220.

von Ihering, H. 1882. On the formation of strata by ants. (In German.) Neues Jahr. Mineral. Geol. Paleontol. 1:156–157.

Vreeken, W.J. 1975. Principal kinds of chronosequences and their significance in soil history. J. Soil Sci. 26:378–394.

Williams, B.H. 1945. Sequence of soil profiles in loess. Am. J. Sci. 243:271–277.

Wright, V.P. (ed.). 1986. Paleosols: Their recognition and interpretation. Princeton Univ. Press, Princeton, NJ.

Yaalon, D.H. 1971. Soil-forming processes in time and space. p. 29–39. *In* D.H. Yaalon, (ed.) Paleopedology: Origin, nature and dating of paleosols. Keter Press, Jerusalem.

Yaalon, D.H. 1983. Climate, time and soil development. p. 233–251. *In* L.P. Wilding et al. (ed.) Pedogenesis and soil taxonomy I: Concepts and interactions. Elsevier Sci., Amsterdam.

8 Towards a New Framework for Modeling the Soil-Landscape Continuum

K. McSweeney
University of Wisconsin
Madison, Wisconsin

B. K. Slater
University of Wisconsin
Madison, Wisconsin

R. David Hammer
University of Missouri
Columbia, Missouri

J. C. Bell
University of Minnesota
St. Paul, Minnesota

P. E. Gessler
CSIRO Division of Soils,
Australian National University
Canberra, Australia

G. W. Petersen
Pennsylvania State University
University Park,
Pennsylvania

ABSTRACT

This paper provides: (i) a brief historical review of soil-geomorphological approaches to investigations of soil-landscape formation, and (ii) an outline of a methodological and conceptual approach for three-dimensional (3-D) modeling of the soil-landscape continuum that utilizes geographic information systems (GIS), spatial analysis and field data. Four interrelated, iterative stages for developing 3-D models of the soil-landscape are outlined. The stages are designed to be explicit and applicable to the scale accuracy specified by the user. The first involves assembly and analysis of pertinent data to characterize a physiographic domain. The second stage is a geomorphometric characterization of the landscape from digital terrain models, which provides (i) a land surface representation to which other data are referenced and (ii) a division of the land surface into areas that correspond with soil patterns. The third

---

Copyright © 1994 Soil Science Society of America, 677 S. Segoe Rd., Madison, WI 53711, USA. *Factors of Soil Formation: A Fiftieth Anniversary Retrospective.* SSSA Special Publication 33.

stage uses georeferenced sampling as a basis for defining soil horizons, their attributes, and spatial arrangement in the landscape. The use of soil horizons rather than pedons as primary extrapolative units is a departure from traditional methods, but is central to developing a 3-D approximation of the soil diversity in the landscape. The fourth stage addresses the basic structure of the model to provide insight about (i) the range, variance, and correlation of soil and associated landform attributes and (ii) the pedogeomorphic processes that formed the landscape. The four stages may be viewed as an integrated procedure for defining explicitly decisions/assumptions, data analysis, visualization, and quantification of the scale and frequency of horizon patterns in three and four (time) dimensional landscapes.

Our major premise is that improved understanding of soil formation must proceed from better approximations of the soil-landscape continuum in scale, space, and time. The conceptual appreciation of this issue has long been recognized in pedology (e.g., Milne, 1935; Jenny, 1941; Ruhe, 1956; Butler, 1959). However, we have lacked the tools for organization and analysis of data to address the tremendous complexity that has resulted in evolution of the mosaic of contemproary soil landscapes. Instead, it has been necessary to rationalize the complexity via taxonomic concepts and mapping procedures that are somewhat subjective and generalized in terms of depicting spatial reality. Although traditional procedures may involve point observations and stratigraphic principles, this information is often lost or obscured in published product because of the rigidity imposed by conventional design of map units.

In many instances, similar taxonomic and mapping principles are used to guide selection of "representative" landforms and associated profiles or pedons for studies of soil formation. However, current interest in soil spatial variability, map unit purity and representation of map unit inclusions, illustrate a need for more quantitative treatment of soil-landform relationships at the mesoscale (Fig. 8-1). This has important implications for selecting points in the landscape (sensu "representative" profiles) from which detailed, time-consuming analyses are made for interpretation of broader-scale processes of soil and landscape formation.

A major challenge for earth scientists is to characterize landscapes into domains where soil, hydrologic and landform attributes can be considered products of common processes of formation and function in an integrated manner (Gerrard,1990). Such a synthesis is desirable to advance fundamental knowledge of earth surface processes operating at a local scale. Ideally, the synthesis also should be suitable for incorporation of process-based models of soil formation and integration with other earth surface and ecological models (Beven, 1989; Christophersen & Neal, 1990). Although the close relationship among soil, landform and hydrology has long been recognized (Milne, 1935, 1936; Gerrard, 1990), a coherent framework that promotes seamless disciplinary integration has yet to emerge.

In this paper, we outline a procedure for developing more realistic models of the soil-landscape continuum. Our model builds on technical advances made through linkage of hydrology and geomorphometry for terrain-based

MODELING THE SOIL-LANDSCAPE CONTINUUM

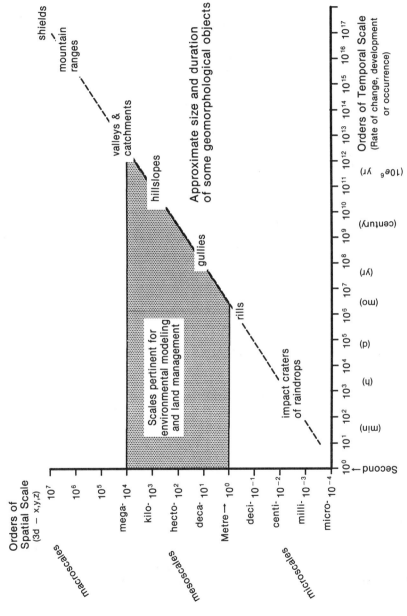

Fig. 8-1. The space-time continuum (modified from Ahnert, 1988; Dikau, 1989).

modeling of hydrological processs (e.g., Moore et al., 1991). The approach uses GIS technology through which various sources of spatial (e.g., vegetation, bedrock geology) and attribute (e.g., soil organic matter content, particle-size distribution) data may be integrated. The data sets may be viewed singly or in combination as representtions of Jenny's (1941) soil-forming factors. Specifically, we will discuss procedures for incorporation of soil horizon stratigraphy into terrain models. We contend that addition of the subsurface dimension into terrain models will provide a platform for investigation of earth surface processes and, more specifically, for refining current understanding of the evolution of soils and landforms.

Our approach addresses largely the mesoscale of the space-time continuum (Fig. 8-1), which is the traditional focus for soil-landscape modeling and its application to land use. However, our approach requires incorporation of pedological data and interpretation of processes operating at microscales (Fig. 8-1). In principle, the approach should be adaptable for addressing macroscale processes operating at regional levels and beyond. The approach also provides a more refined and precise framework for depicting and interpreting the soil-landscape continuum that can be linked with pedological process models (e.g., hydrology, biogeochemistry) more easily than from conventional soil maps.

The objectives of the paper are to: (i) provide a brief historical review of soil-geomorphological approaches to investigations of soil formation, and (ii) illustrate an approach for 3-D modeling of the soil-landscape continuum that uses the organizational and spatial analytical framework provided by GIS and allied technology.

HISTORICAL PERSPECTIVE

The early development of pedology (1870s-1940s) was based largely on investigations of soils in the northern hemisphere developed under limited geological time, from relatively uniform glacially derived materials. Several important, but not necessarily consistent themes, emerged from these investigations.

The biogeographic concept of soil zonality strongly influenced the major soil classification schemes developed during this period (Baldwin et al., 1938; Gerasimov et al., 1939, as reviewed by Buol et al., 1989). The "zonal" or "mature" soil concept represented a dynamic steady-state alteration of parent material adjusted to the environment (Nikiforoff, 1942). This concept implied a constancy in erosion and deposition rates and placed a major emphasis on the interdependent influences of contemporary climate and vegetation on soil maturation. Butler (1964) and Firman (1968) suggested that the "zonality" concept stifled development of the stratigraphic significance of soils.

Formalization of the factors of soil formation (Jenny, 1941), the roots of which can be traced to Dokuchaev (Tandarich & Sprecher, 1991), fostered the concepts of soils as integrated components of ecosystems. Jenny's incorporation of topography as a factor provided some accommodation between geomorphology and pedology, but specific investigations of soil-landform relationships were only nascent. For example, Milne's (1935, 1936) pioneering work in East Africa spawned the catena concept, which emphasized integrated relationships among slope/landform and hydrology/morphology, though with a strong geological bias. Paleopedology, and by implication recognition of episodic development of soil landscapes, had received some attention (Polynov, 1927; Thorp, 1935; Albritton & Bryan, 1939), but does not appear to have influenced mainstream pedology until after World War II.

In the 1950s, a systematic framework for investigation of soil-landscape relationships emerged (Ruhe, 1956; Butler, 1959). The Ruhe/Butler approach is based on geomorphic and stratigraphic principles and explicitly recognizes the periodic nature of soil-landscape development. This approach has strongly influenced many subsequent soil-landscape studies and has improved qualitative understanding of soil, landform and stratigraphic relationships (Olson, 1989).

In routine soil survey attributes of soil map units are referenced to modal pedons or profiles designed to suit the study area, which in turn are referenced to nationally or internationally recognized classification schemes. Thus, information about internal variation within map units and stratigraphic relationships among map units is limited (Fig. 8-2). This approach has served the largely utilitarian goals of national soil survey programs admirably, but has limitations for portraying soil information suitable for contributing to integrated investigations of earth surface processes at the local scale. It is important to recognize that conventional soil survey was well established prior to emergence of the tools and techniques for more sophisticated analysis and portrayal of geographic information.

CONTEMPORARY PERSPECTIVE

A pressing challenge is to establish relationships between soils and landforms at large scales and relate the patterns to processes of pedogeomorphic evolution. Gerrard (1990) demonstrates by reference to several studies conducted in the United Kingdom that not all slopes possess fully integrated soil systems. Different parts of the slopes may act independently. Such discordance is likely in landscapes composed of a variety of parent materials that have been subjected to climate, land use, and pedogeomorphic change. An important question that needs to be considered in development of soil landscape models is: How strong are the relationships among soils and landforms across the landscape and at what scales are the relationships most evident?

The landscape, like the soil, is a continuum. Scale dependence complicates the definition of simple slope units with their relationships to soil attributes. Even sophisticated conceptual land-surface models (Dalrymple et al., 1968; Conacher & Dalrymple, 1977) can impose an arbitrary division of the continuum. No single model can accommodate observed variability operating at the local scale (Gerrard, 1990). Conceptual landform models

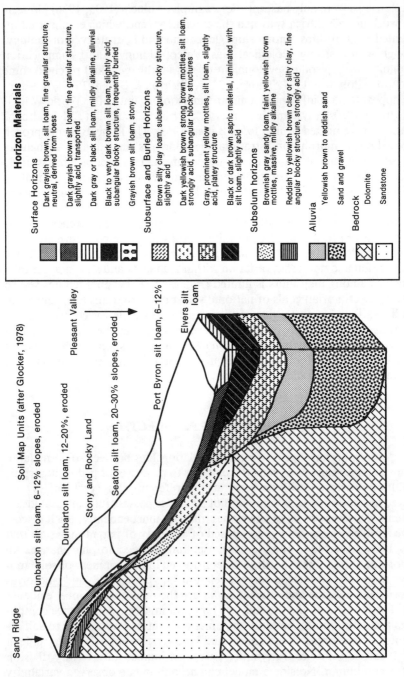

Fig. 8-2. Cross-section illustrating complexity of major horizons present along transect from Sand Ridge to Pleasant Valley, WI. The legend lists preliminary horizons characterized by simple morphological attributes (Glocker, 1978).

appear limited to unraveling the 3-D complexity inherent in soil–landform relationships. Limitations also result from the rigidity imposed by linkage of field observations via modal concepts to soil classification schemes based on idealized universes. In practice, this problem is often confounded through attempts to correlate defined slope segments with soil attributes sampled from a linear transect—in essence, using a two-dimensional (2-D) design to address a 3-D problem.

PROPOSED FRAMEWORK FOR DEVELOPING A SOIL–LANDSCAPE MODEL

Model Organization

Four interrelated, iterative stages for developing a soil–landscape model (Fig. 8-3) are summarized below. The stages are explained in greater detail in later sections. The approach is iterative, and uses a combination of spatial analysis and field data to develop an explicit model that may be applied to the scale accuracy level specified by the user.

The first stage involves integration of available data sets (e.g., geology, topography, climate past and present, vegetation, remotely sensed data) to define and characterize the physiographic area under study, to consolidate our a priori knowledge about the area, and to identify other data that might be valuable for defining soil patterns. The data sets may be viewed, singly or in combination, as representations of Jenny's soil-forming factors. For example, Gessler et al. (1989) have used information about presettlement (1830s) vegetation and subsequent historical changes in vegetation to reconstruct temporal changes in soil in response to changes in land cover and use in the Driftless Region of southwestern Wisconsin. The second stage is a geomorphometric characterization of the landscape by primary and secondary landscape attributes (Table 8-1) derived from a digital elevation model (DEM) (e.g., Moore et al., 1991). Primary attributes are directly calculated from elevation data and include the first and second vertical and horizontal derivatives (slope, profile curvature; aspect, plan curvature) and flow direction. Secondary attributes are simple or complex combinations of primary attributes and may provide indices that describe or characterize the spatial variability of specific landscape processes such as variation in soil moisture content and erosion potential (Moore et al., 1991). The DEM is then used for (i) representing the land surface to which all point information is subsequently referenced, and (ii) geomorphometric characterization via primary and secondary attributes of the land surface into areas that may correspond with soil patterns. In this sense, the model places a special emphasis on quantifying Jenny's relief or topographic factor, which is particularly important for large-scale investigation of soil patterns (Jenny, 1941).

The third stage involves development of a soil horizon legend that is used to determine the distribution and spatial relationship among soil horizons and other layers in the landscape. Horizons have been chosen in preference

Table 8-1. Primary and secondary attributes derived from digital terrain model (DTM) analysis (adapted from Speight, 1977; Moore et al., 1991, 1993).

Attribute	Definition	Significance
\multicolumn{3}{c}{Primary attributes}		
Altitude	Elevation	Climate, vegetation, potential energy
Aspect	Slope azimuth	Solar insulation, evapotranspiration, flora and fauna distribution and abundance
Slope	Gradient	Overland and subsurface flow velocity and runoff rate, precipitation, vegetation, geomorphology, soil water content, land capability class
Profile curvature	Slope profile curvature	Flow acceleration, erosion/deposition rate, geomorphology
Plan curvature	Contour curvature	Converging, diverging flow, soil water content, soil characteristics
\multicolumn{3}{c}{Secondary attributes}		
Upslope slope	Mean slope of upslope area	Runoff velocity
Dispersal slope	Mean slope of dispersal area	Rate of soil drainage
Catchment slope	Average slope over catchment	Time of concentration
Upslope height	Mean height of upslope area	Potential energy
Upslope area	Catchment area above a length of contour	Runoff volume, steady-state runoff rate
Dispersal area	Area downslope from a short length of contour	Soil drainage rate
Catchment area	Area draining to catchment outlet	Runoff volume
Specific catchment area	Upslope area per unit width of contour	Runoff volume, steady-state runoff rate, soil characteristicfs, soil water content, geomorphology
Flow path length	Maximum distance of water to a point in the catchment	Erosion rates, sediment yield, time of concentration
Upslope length	Mean length of flow paths in the catchment	Flow acceleration, erosion rates
Dispersal length	Distance from a point in the catchment to the outlet	Impedance of soil drainage
Catchment length	Distance from highest point to outlet	Overland flow attenuation
Wetness index	Computed using specific catchment area and slope	Soil moisture characteristics
Streampower index	Computed using specific catchment area and slope	Erosive power of overland flow
Sediment transport capacity indexd	Computed using specific catchment area and slope	Erosion and deposition processes
Solar radiation indices	Computed using terrain, climate, surface reflectance data	Energy availability and flux

to profiles as the basic unit for field investigation because they provide a more convenient means for explicit representation of the soil landscape. This stage requires field investigation in which all information and samples collected for laboratory analysis are georeferenced. The fourth stage involves laboratory and statistical analyses of soil horizon attributes collected during the third stage. This stage serves largely to refine (i) the horizon stratigraphy and its relationship to geomorphometrically defined landscape patterns, and (ii) our pedological and geomorphological understanding of the area.

The four stages proceed from coarser to finer scales of analysis and provide a hierarchy for quantifying and modeling soil-landscape patterns at various scales and frequencies. Hierarchy theory (Allen & Starr, 1982) proposes that natural processes operate at various spatiotemporal scales and frequencies. Scale dependency is an essential feature of soil attributes and processes (Kachanoski, 1988), which we attempt to accommodate in our model.

Within each stage (e.g., horizon stratigraphy characterization) steps of data gathering and input, model development and data analysis, and testing, refining and output are linked. The diamonds found in the horizontal rows of each stage (Fig. 8-3) symbolize steps in model development that require explicit definition of assumptions or decisions before the next step can proceed. These steps enforce definition of decisions such as types of data structure used for modeling, scale or frequency of sampling, method of measurement and associated representative elementary volume, selection of predictive equations, and statistical interpolation and visualization methods. All such decisions are premised on assumptions that need to be defined so that refinement of the model can proceed as improved knowledge of the area is acquired. These steps are designed to reduce subjective assessment and to serve the needs of model testing and refinement.

An important assumption in our approach is that soil and geomorphic patterns covary and are linked to process. Thus, areas should exist on the landscape where there is a strong correlation between landform and the underlying nature and arrangement of soil horizons. The challenge is to: (i) identify and define where landform-soil horizon relationships are strong, (ii) determine the feasibility of using these relationships for extrapolation and correlation across the landscape, and (iii) interpret these relationships in terms of process and events that result in soil-landscape evolution. For example, eolian-derived silt on a hillslope may show fairly consistent trends of increasing thickness and change in nature and arrangement of soil horizons from summit to footslope that is related to initial deposition and subsequent pedological and geomorphic processes. If the soil pattern conforms to a spatial pattern defined by relative elevation and slope geometry, it may be feasible to extrapolate and correlate these relationships more broadly across the landscape.

Nevertheless, some physiographic domains may contain areas of very complex landform-soil horizon patterns (e.g., in areas with cradle-knoll or gilgai microtopography), which occur over such short distances that special treatment will be necessary for their portrayal in a 3-D model. For example, areas may occur that cannot be reasonably resolved with existing assump-

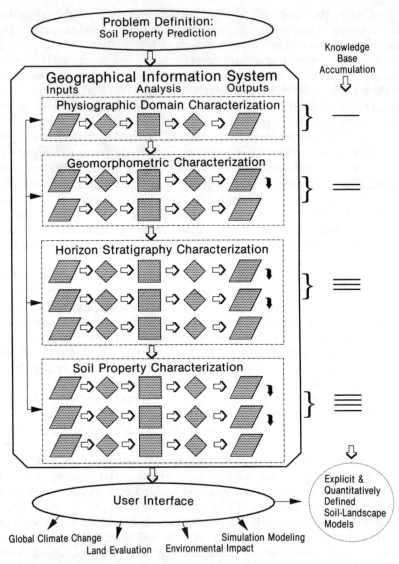

Fig. 8-3. Scheme for development of soil–landscape models.

tions and techniques. Landscapes in which subsurface features (e.g., variation in bedrock geology) or process (e.g., saline groundwater) exert a strong influence on the contemporary soil pattern may pose particular challenges for resolving landform–soil horizon patterns. However, the operational framework allows incorporation of new data and assumptions required for refinement of the model.

The approach integrates GIS, image processing, and statistical analysis software. Since no single system exists to meet all requirements, users must mix and match software and hardware platforms of choice. As such, the approach outlined in this paper focuses on a generic framework afforded by an integration of an array of spatial analysis techniques rather than a detailed exposition of the technology itself.

Physiographic Domain Characterization

Quantitative soil information is required for understanding the structure and function of ecosystems over a range of scales (Jenny, 1980; Turner & Gardner, 1990). Integrated disciplines such as landscape ecology (Naveh & Lieberman, 1984) and the data needs of land evaluation and environmental impact assessment require that we explicitly state where a particular soil–landscape model may or may not apply.

The first step in the physiographic domain characterization (Fig. 8-3) is to input pertinent data sets that may differentiate the domain(s) in which the model may apply. Scale and source of each data set must be defined, because this information is essential in any decisions or assumptions for subsequent use of the data. Analyses may then proceed by spatial overlay to create combinations of attributes (e.g., geology, climate, vegetation) characterizing the domain with careful attention to scale compatibilities. The specificity of the characterization will depend on the quality of the existing data sets and can be improved as additional information becomes available. The output of the first submodule is a broad characterization of the physiographic macrodomain (Dikau, 1989) under study, which together with any a priori knowledge, provides cursory understanding of the relationship between factors of soil formation and soil patterns. Subsequent substages (Fig. 8-3) lead to a further breakdown into mesodomain process zones within the macrodomain to provide the basis for sampling and modeling the soil–landscape continuum.

Geomorphometric Characterization

The foundation for modern geomorphometric analysis was established by Strahler (1952), Ruhe and Walker (1968), and Speight (1968), who lacked the tools for practical implication of quantitative concepts at the level of sophistication available today. Evans (1972) and Lanyon and Hall (1983) described quantitative techniques and recently a quantitative framework using digital terrain model data has been implemented by Dikau (1989). Additional terrain attributes describing basic physical processes have been outlined by Moore et al. (1991, 1993).

Digital elevation models provide a representation of the external landscape continuum to which horizon stratigraphy can be added. Geomorphic landforms and horizon patterns are often strongly correlated because of common formative processes. Traditionally, these correlations, as observed at points or transects, have been extrapolated using subjective air photograph interpretation. The result is a delineation of implied, but perhaps unintended, homogeneous units with unknown internal variance.

Digital terrain analysis methods can provide explicit, efficient, and consistent quantitative techniques for geomorphometric characterization. Additionally, terrain analysis provides a mechanism for incorporating spatial attributes that describe the basic physics of earth surface and subsurface processes. Table 8-1 lists primary and secondary attributes derived from a DEM and their definition and significance with respect to landscape processes. Such attributes can be used to quantitatively describe water, solute and material flow pathways, and the contextual nature of a specific point in a catchment or watershed. Since horizon differentiation is primarily driven by energy and material fluxes (Simonson, 1959; Smeck et al., 1983), such attributes are useful for describing soil formation processes and related horizon patterns. Hence, we no longer need to impose potentially inappropriate and static slope classes onto the landscape, and can model simple or complex attribute overlays and aggregations as a continuum. Likewise, the attributes may be used to spatially predict finer-scale process-based soil attribute patterns. For example, Moore et al. (1993) demonstrate the use of terrain attributes for modeling such soil attributes as depth of A horizon, pH, extractable P, soil texture, and organic matter levels in the surface horizon of a toposequence in Colorado. Figure 8-4 illustrates depth of A horizon as modeled by traditional soil survey (Fig. 8-4a) and as measured at 231 points on a 15-m grid over this toposequence (Fig. 8-4b). Figure 8-4c and 8-4d show primary (slope) and secondary (wetness index) terrain attributes with computed correlation coefficients of 0.55 and 0.64 with measured A-horizon depth (Fig. 8-4b). This illustrates, simply, quantitative terrain attributes that relate to both physical processes (e.g., sediment transport, water availability, energy flux) and patterns of the soil–landscape continuum. Geographic information systems, image analysis, and statistical techniques provide a diverse tool set for visualizing patterns and relationships in 2-, 3- or four-dimensions (4-D) (time).

Two stages are proposed within the geomorphometric submodule (Fig. 8-3). The first requires the input of a DEM for the generation and output of primary and secondary terrain attributes (Table 8-1). The second inputs these derivatives, along with other previously gathered data sets, for spatial exploratory data analysis. These data can be used to define patterns and characterize the external landform to assist in the development of efficient sampling strategies within the domain of study (Fig. 8-3).

Various methods and sources for DEM development and input exist (e.g., via global positioning systems, photogrammetry, contours) and will not be expounded here. Decisions are required for selection of the data structure (e.g., grid, contour) and scale or frequency of DEM sampling. Ahnert (1988) and Dikau (1989) provide useful discussion and illustration of size orders of relief—Fig. 8-1 is adapted from this work. Resolution of particular geomorphological objects and landscape morphological features using primary and secondary terrain attributes will depend on the sampling frequency (i.e., grid spacing) and data structure. Further decisions for the computational algorithms to derive the descriptive terrain attributes follow once a data structure and sampling frequency are chosen. After these decisions are documented, primary and secondary terrain attributes are computed.

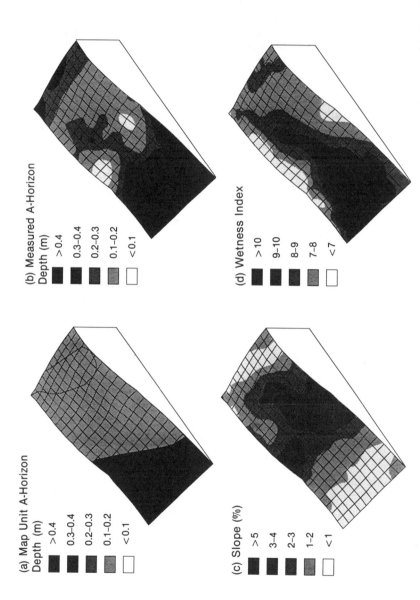

Fig. 8-4. "A" horizon thickness derived from soil survey (*a*), from grid sampling (*b*), and slope (*c*), and wetness index (*d*) derived from digital terrain model analysis for a toposequence in Colorado (after Moore et al., 1993).

The output of the geomorphic characterization provides an enhanced spatial data set with defined patterns to be used for hypothesis testing through the next substage of horizon characterization. What formerly ended with the subjective and qualitative delineation of a pattern on an air photograph can now be explicitly defined for testing, validation and extrapolation. However, a definitive methodology for geomorphometric characterization does not exist and requires considerable empirical research.

Horizon Characterization

Soil horizons are a result of geomorphic and pedogenic processes (Simonson, 1959). A framework for study of their "stratigraphic" arrangement was outlined by Butler (1959, 1982), and furthered by Vreeken (1984). The horizon has been used for soil classification (FitzPatrick, 1967; Soil Survey Staff, 1975) and was recently proposed as a key "carrier" of information (Bouma, 1989). The time may be ripe for a new scientific paradigm using soil horizon entities to model the soil continuum.

Terrain analysis provides a framework from which modeling of underlying horizon strata may proceed. However, because we cannot see the interwoven nature of the horizon strata, we must rely either on point/trench samples and correlations with simpler derived attributes (i.e., geomorphometric) for extrapolation or new methods of scanning stratified soil materials. Three linked steps are proposed within this substage (Fig. 8-3). The first involves a gathering of available horizon data (e.g., soil maps) and/or a reconnaissance sampling to develop a horizon legend and initial model. The second step evaluates the correlation of the internal horizon stratigraphy with external landform. The third step solidifies an understanding of stratigraphic patterns through variable-scale nested sampling and trenches to refine and test the model.

The inputs to the first step are existing soil maps (if available) or, preferably, the original sample points and a reconnaissance sampling planned with knowledge gained in previous modules. Decisions required include a sampling strategy and a defined methodology for explicitly defining a horizon or layer. Sampling may be planned in numerous ways, including an informed forward hypothesis-testing fashion where patterns derived in prior submodules can be explored. Alternatively, point samples can be used in a "truthing" fashion to search spatial data sets for correlation and predictive/extrapolative potential. Sampling also may require iterative collection and evaluation to ensure that all important mesodomain horizon entities are incorporated into the soil–landscape model.

Inputs to the second step include previously collected landscape and environmental attributes, and data from the previous step defining the spatial occurrence of horizons. Correlation analysis may then proceed to decipher the quantitative relationship between external landform and internal soil horization. Decisions about methodology must be documented so that others may test any developed model. If initial analysis is inconslusive, additional sampling may be required. The output of the second step is a soil–landscape model defined by explicit quantification of correlations between horizon stratigraphic patterns and other existing attributes. The model should provide a probabilistic estimate for the occurrence and size dimension of a particular horizon at a point. If correlations do not exist, then the only way to model horizon stratigraphy is by sampling. Similarly, if correlation is strong in defined parts of the domain but not in others (Gerrard, 1990), this provides information about where best to expend expensive field resources.

The final step tests the developed soil-landscape model, Inputs include the model itself and sample points not used for model development. Analysis is carried out to evaluate the predictive and extrapolative potential within the domain of study and also to rate the rigor of the model as scale of sampling is varied. This may best be done using a geostatistical design based on accumulated information. It is important to determine the scale of correlate sampling needed to meet the predictive specifications set by the user. Each georeferenced point provides information not only about a horizon's vertical size and locational arrangement with respect to other horizons, but also references the soil surface and the rest of the physiographic domain. Spatial searches for all horizons between certain relative elevations on particular parent materials may lead to development of complex decision sets relating geochemistry and hydrology to pedological and other earth surface and subsurface processes. Relationships among external landform, soil horizonation, and the consolidated/unconsolidated material contact also should be determined to elucidate the entire range of contributing factors to soil-landscape patterns.

A seminal paper by Butler (1964) outlined that, because the spatial scale of variation differs from one soil property to another, taxonomic soil entities and patterns differentiated according to morphology may be inadequate for modeling the variation of other physical, chemical, and biological properties. Even though developed relationships may be locale specific (i.e., a specific physiographic domain), they are nevertheless important for extension of research and soil technology. Regionalized variable theory or geostatistics was in its infancy at this time (Matheron, 1962, 1963a,b) and spatial statistical methods and applications have advanced considerably (Webster & Oliver, 1990; Cressie, 1991). However, Butler's arguments remain significant and must be cast in the light of current model development. Many of today's methods have not been applied in an integrated manner to relate pattern to process for development of holistic models of the soil-landscape system.

Soil Property Characterization

Users require data describing the variation/pattern of a particular soil property over a spatial domain for simulation modeling, land evaluation, environmental impact assessment and so on. The objective of the final substage (Fig. 8-4) is to determine the most efficient means for achieving specified data requirements. The first of three steps requires an initial property sampling within the major stratigraphic horizons to determine if variation is less within horizons than between horizons. Second, geostatistically designed sampling strategies may be employed to estimate the ranges, variances, covariances, and cross-correlations of various properties (e.g., potential surrogates) and respective spatial correlation structures. The final step involves an integration of acquired knowledge and quantitative relationships to specify the most efficient means for spatial prediction or modeling the property of interest.

The sampling strategy employed and number of samples collected in the first step must ensure a level of confidence that the within- and between-horizon variation is adequately described. Additional decisions about sampling methods are required (i.e., sample, volume, laboratory techniques, temporal frequency). The first step yields an indication of the usefulness of horizons for modeling the soil properties of interest and the potential modeling expense if horizons are not shown to be valid "carriers" of information. It also will suggest if lumping of horizons is practical for specified properties.

The second step is used to determine if the statistical results of the first step also hold true in a spatial sense. Decisions are needed for a scale-variable nested sampling strategy to define spatial correlation structures (variograms). Does systematic spatial variation exist relating to particular process frequencies? Likewise, if one or a combination of properties are proposed as surrogates for a more difficult to measure property, it is important to know the spatial correlation structures of each and how detected differences may influence spatial prediction error.

Finally, the systematic gathering of data and development of quantitative relationships culminates in the synthesis of a knowledge base for definition of an explicitly defined soil–landscape model. As significant as the model itself, the framework provides an open forum for independent interdisciplinary testing and refinement so that researchers with particular expertise can contribute to evolution of better soil–landscape continuum models. Clearly, the final phase of model development is essential for refining the utility of the horizon stratigraphy developed in the previous submodule, and may result in new configurations of the stratigraphy to address specific uses of the model. For example, the model may utilize groupings of subsurface horizons based on threshold hydraulic conductivities for modeling lateral flow, whereas agronomic applications may require subdivision of surface horizons to address spatial variation in nutrients within a plow layer that is stratigraphically defined as a distinct entity.

SUMMARY

In summary, the matrix (Fig. 8–3) is designed to formalize the definition of the stage used in developing a soil–landscape model. Each stage and substage may be tested and improved via feedback and revision. This requires explicit decision definitions about sampling methods, data structures, data analysis, visualization, and quantification of the scale and frequency of horizon stratigraphic and soil attribute patterns in 3- and 4-D (time) landscapes. This holistic methodology facilitates the systematic development of a knowledge base to aid spatial prediction and our basic understanding of landscape processes.

For a long time, basic correlations have been assumed but not tested. The framework outlined provides an initial attempt at the development of a comprehensive methodology for testing and incorporating quantitative relationships into spatially rigorous soil–landscape models. Such models are crucial if we are to advance our basic understanding of earth surface and subsurface processes.

Geographic information systems and allied spatial analytical techniques provide tools for quantitative analysis of relationships between landform and soil attributes. A systematic framework for soil/landform landscape structure can be developed by iterative correlation of field-measured soil attributes arranged by horizon with geomorphic attributes and domains derived by quantitative terrain analysis. Resultant soil–landscape models will provide the basis for extrapolation on a broader scale and for more reliable interpretations of earth history and land use. The approach draws strongly on the factorial approach formalized by Jenny (1941) and attempts to embellish it through quantitative spatial analysis and emphasis on elucidating pedogeomorphic processes.

ACKNOWLEDGMENTS

Kevin McSweeney's contributions to this paper were supported in part by a grant from the National Science Foundation (SES-9210093) and those of Paul Gessler by a grant from the Murray-Darling Basin Commission (NRMS-M218). We thank Ron Amundsen, Jennifer Hardin, and Mike Singer for their constructive comments on an earlier draft of the manuscript.

REFERENCES

Ahnert, F. 1988. Modelling landform change. p. 375–400. *In* M.G. Anderson (ed.) Modelling geomorphological systems. Catena Suppl. 6. Catena Verlag Publ., Cremingen-Destadt, the Netherlands.

Albritton, C.C., Jr., and K. Bryan. 1939. Quaternary stratigraphy in the Davis Mountains, Trans-Pecos, Texas. Bull. Geol. Soc. Am. 50:1423–74.

Allen, T.F.H., and T.B. Starr. 1982. Hierarchy: Perspectives for ecological complexity. Univ. of Chicago Press, Chicago.

Baldwin, M., C.E. Kellog, and J. Thorp. 1938. Soil classification. p. 979–1001. *In* Yearbook of agriculture. USDA, U.S. Gov. Print. Office, Washington, DC.

Beven, K. 1989. Changing ideas in hydrology—The case of physically-based models. J. Hydrol. 105:157–172.

Bouma, J. 1989. Land qualities in space and time. p. 3–13. *In* J. Bouma and A.K. Bregt (ed.) Land qualities in space and time. Pudoc, Wageningen, the Netherlands.

Buol, S.W., F.D. Hole, and R.J. McCracken. 1989. Soil genesis and classification. 3rd ed. Univ. Press, Ames, IA.

Butler, B.E. 1959. Periodic phenomena in landscapes as a basis for soil studies. Soil Publ. 14. CSIRO, Div. of Soils, Canberra, Australia.

Butler, B.E. 1964. Can pedology be rationalized: A review of the general study of soils. p. 3. *In* Presidential address. Publ. 3. Australian Soc. Soil Sci., CSIRO, Div. of Soils, Canberra.

Butler, B.E. 1982. A new system for soil studies. J. Soil Sci. 33:581-595.

Christophersen, N., and C. Neal. 1990. Linking hydrological, geochemical, and soil chemical processes on the catchment scale: An interplay between modeling and field work. Water Resour. Res. 26:3077-3086.

Conacher, A.J., and J.B. Dalrymple. 1977. The nine unit landsurface model: An approach to pedogeomorphic research. Geoderma 18:1-154.

Cressie, N.A. 1991. Statistics for spatial data. John Wiley & Sons, New York.

Dalrymple, J.B., R.J. Blong, and A.J. Conacher. 1968. A hypothetical nine unit landsurface model. Z. Geomorph. 12:60-76.

Dikau, R. 1989. The application of a digital relief model to landform analysis in geomorphology. p. 51-79. *In* J. Raper (ed.) Three dimensional applications in geographical information systems. Tayler & Francis, London.

Evans, I.S. 1972. General geomorphometry, derivatives of altitude, and descriptive statistics. p. 17-90. *In* R.J. Chorley (ed.) Spatial analysis in geomorphology. Methuene, London.

Firman, J. 1968. Soil distribution—a stratigraphic approach. p. 569-576. *In* Proc. Int. Congr. Soil Sci., 8th, Adelaide. Vol. 4. Am. Elsevier Publ. Co., New York.

FitzPatrick, E.A. 1967. Soil nomenclature and classification. Geoderma 1:91-105.

Gerasimov, I.P., A.A. Zavalishin, and E.N. Ivanova. 1939. A new scheme of a general soil classification of the USSR (In Russian.) Pochvovedenie 7:10-43.

Gerrard, A.J. 1990. Soil variations on hillslopes in humid temperate climates. Geomorphology 3:225-244.

Gessler, P.E., K. McSweeney, R. Kiefer, and L. Morrison. 1989. Analysis of contemporary and historical soil/vegetation/land use patterns in southwest Wisconsin utilizing GIS and remote sensing technologies. p. 85-92. *In* Tech. Pap. 1989. Am. Soc. of Photogram. and Remote Sensing and Am. Congr. Surveying and Mapping, ASPRS, CSM, Falls Church, VA.

Glocker, C.L. 1978. Soil survey of Dane County, Wisconsin. USDA-SCS.

Jenny, H. 1941. Factors of soil formation: A system of quantitative pedology. McGraw-Hill, New York.

Jenny, H. 1980. The soil resource; Origin and behavior. Ecol. Studies 37. Springer-Verlag, New York.

Kachanoski, R.G. 1988. Processes in soils—from pedon to landscape. p. 153-177. *In* T. Rosswall et al. (ed.) Scales and global change. John Wiley & Sons, New York.

Lanyon, L.E., and G.F. Hall. 1983. Land-surface morphology: 1. Evaluation of a small drainage basin in eastern Ohio. Soil Sci. 136:291-299.

Matheron, G. 1962. Traite de geostastique appliquee, tome I. memoires du bureau de recherches geologiques et minieres. No. 14. Editions Technip, Paris.

Matheron, G. 1963a. Traite de geostatistique appliquee, tome II, le krigeage. memoires du bureau de recherches geologiques et minieres. No. 24. Editions Bureau de recherches geologiques et minieres, Paris.

Matheron, G. 1963b. Principles of geostatistics. Econ. Geol. 58:1246-1266.

Milne, G. 1935. Some suggested units of classification and mapping particularly for East African soils. Soil Res. 4:3.

Milne, G. 1936. Normal erosion as a factor in soil profile development. Nature (London) 138:548-549.

Moore, I.D., R.B. Grayson, and A.R. Ladson. 1991. Digital terrain modelling: A review of hydrological, geomorphological, and biological applications. Hydrol. Proc. 5:3-30.

Moore, I.D., P.E. Gessler, G.A. Nielsen, and G.A. Peterson. 1993. Soil attribute prediction using terrain analysis. Soil Sci. Soc. Am. J. 57:443-452.

Naveh, Z., and A.S. Lieberman. 1984. Landscape ecology: Theory and application. Springer-Verlag, New York.

Nikiforoff, C.C. 1942. Fundamental formula of soil formation. Am. J. Sci. 240:847-66.

Olson, C.G. 1989. Soil geomorphic research and the importance of paleosol stratigraphy to Quaternary investigations, Midwestern USA. p. 129-142. *In* A. Bronger and J. Catt (ed.) Paleopedology: Nature and application of paleosols. Catena Suppl. 16. Catena Verlag Publ., Cremlingen-Destedt, the Netherlands.

Polynov, B.B. 1927. Contributions of Russian scientists to paleopedology. Russian Pedological Investigations, Part VIII. Acad. Sci., Moscow, USSR.

Ruhe, R.V. 1956. Geomorphic surfaces and the nature of soils. Soil Sci. 82:441-55.

Ruhe, R.V., and P.H. Walker. 1968. Hillslope models and soil formation: I. Open systems. p. 551-560. *In* Proc. Int. Congr. Soil Sci. Soc., 9th, Adelaide. Vol. 4. Elsevier Publ. Co., New York.

Simonson, R.W. 1959. Outline of a generalized theory of soil genesis. Soil Sci. Soc. Am. Proc. 23:152-156.

Smeck, N.E., E.C.A. Runge, E.E. Mackintosh. 1983. Dynamics and genetics modeling of soil systems. p. 51-81. *In* L.D. Wilding et al. (ed.) Pedogenesis and soil taxonomy. I. Concepts and interactions. Elsevier, Amsterdam, the Netherlands.

Soil Survey Staff. 1975. Soil taxonomy: A basic system of soil classification for making and interpreting soil surveys. USDA-SCS Agric. Handb. 436. U.S. Gov. Print. Office, Washington, DC.

Speight, J.G. 1968. Parametric description of landform. p. 239-250. *In* G.A. Stewart (ed.) Land evaluation. Macmillan Co., Melbourne, Australia.

Speight, J.G. 1977. Land form pattern description from aerial photographs. Photogrammetrica 32:161-182.

Strahler, A.N. 1952. Dynamic basis of geomorphology. Bull. Geol. Soc. Am. 63:923-938.

Tandarich, J.P., and S.W. Sprecher. 1991. The intellectual background for the factors of soil formation. p. 322. *In* Agronomy abstracts. ASA, Madison, WI.

Thorp, J. 1935. Geographic distribution of the important soils of China. Bull. Geol. Soc. China 14:119-46.

Turner, M.G., and R.H. Gardner. 1990. Quantitative methods in landscape ecology: An introduction. p. 3-14. *In* M.G. Turner and R.H. Gardner (ed.) Quantitative methods in landscape ecology: The analysis and interpretation of landscape heterogeneity. Ecol. Stud. 82. Springer-Verlag, New York.

Vreeken, W.J. 1984. Soil-landscape chronograms for pedochronological analysis. Geoderma 34:149-164.

Webster, R., and M.A. Oliver. 1990. Statistical methods in soil and land resource survey. Oxford Univ. Press, Oxford, England.

APPENDIX 1
Memories of Professor Hans Jenny

Rodney J. Arkley
*Professor Emeritus,
Department of Soils and Plant Nutrition
University of California at Berkeley
Berkeley, California*

My first contact with Dr. Jenny was as a student in the first course which he taught in Berkeley—soil colloids. Since he did not have time to prepare a syllabus, he put all the formulae and diagrams on the blackboard. He presented the information very rapidly and managed to cover the blackboard the full width of the lecture room about three times each lecture. There also was no course at the University of California at Berkeley in colloidal chemistry, so one-half of the class consisted of graduate students and seniors from the College of Chemistry. Consequently, we poor soil science majors had to study very hard to obtain decent grades.

Later, I took Dr. Jenny's course in soil genesis using his book *Factors of Soil Formation* as a text. Later, as a soil surveyor for the U.S. Soil Conservation Service in Wisconsin and Illinois, and the University of California, I found that the insight which his ideas provided was invaluable in preparing soil maps. This was particularly important in the selection of soil sampling sites which were representative of each section of the landscape.

I owe a great deal of my career in soil science to Dr. Jenny. He guided my studies toward a Ph.D. degree with a critical but always helpful way. Later we worked together on his favorite project, the Pygmy Forest on the Mendocino, California, coast. What impressed me most about his work was the always meticulous planning and execution of his research, and his boundless enthsiasm for the study of soils and their role as an important factor in ecology. This led to his great interest in preserving areas of soils with special ecological relations. One of the last things he asked me to do, while I was in Russia, was to contact the directors of the Soil Institute of Moscow. Dr. Jenny urged the directors to use the money, accrued to him from the sale of his book in Russia, to purchase and preserve the land around the site where the distinctive Podsol soil was first identified and named.

Copyright © 1994 Soil Science Society of America, 677 S. Segoe Rd., Madison, WI 53711, USA. *Factors of Soil Formation: A Fiftieth Anniversary Retrospective.* SSSA Special Publication 33.

APPENDIX 2
We Remember Hans Jenny

Gordon L. Huntington
University of California
Davis, California

During his lifetime, Hans Jenny left many indelible impressions on his students, friends, and associates. His perception, appreciation, depth of understanding and love of soil often was revealed and focused to enrich contact with this gifted individual. For Jenny, soil was a worldwide tapestry of life forms interwoven with lifeless material in complex, earth-surface systems. Such contact with him often provided stimulus for new thoughts, ideas, or directions for further study among those fortunate enough to know him.

The following vignette provides a glimpse of the influence Hans Jenny has had on those in the field of soil science (aside from his publications).

It was late in the field season of 1958—but springtime in the High Sierra of California. Nearing completion of the Soil-Vegetation Survey of the Sierra National Forest in Fresno County, Jack R. Fisher (U.S. For. Serv.) and Gordon L. Huntington (University of California, Berkeley) were mapping the eastern part of the Kings River District on horseback with pack train and packer. This section was later to become part of the John Muir Wilderness. Three traverses, each a week's duration, were underway.

Jenny had been invited to participate in this prioneering study of High Sierra soils, and had opted for the traverse during the 2nd wk of the work. In preparation for the high country, he took long hikes in the Berkeley Hills during the weeks prior to joining the survey party in Fresno on August 31. On that day, the group traveled to the Crabtree Pack Station, at an elevation of about 1969.7 m (6500 ft), on the North Fork of the Kings River near the newly completed Wishon Dam.

FIRST DAY—WOODCHUCK COUNTRY

Early the following morning, the packer readied the pack train and saddle horses for the survey party. The packer, pack train, and survey party then headed east across Wishon Dam to the westerly slopes of the North Fork canyon. There they took the trail to Crown Lake through the roadless Woodchuck Country. For the first time in the survey, soils formed in glacial ground moraine were encountered under a cover of white fir, ponderosa pine, incense cedar, and sugar pine. The brown color of exposed soil profiles along the trail was evident to all and comparable to the color of the Shaver series formed in the residuum of granitic rock under similar vegetation at similar and lower elevations elsewhere. These morainal soils contrasted with other paler, grayer morainal soils of the Maxson and Dinkey series already recognized and mapped in the area at higher elevations. Quick to perceive the difference, Jenny queried

Copyright © 1994 Soil Science Society of America, 677 S. Segoe Rd., Madison, WI 53711, USA. *Factors of Soil Formation: A Fiftieth Anniversary Retrospective.* SSSA Special Publication 33.

the survey party as to the possibility of a morainal soil climosequence of greater elevational range on the western slopes of the Sierra. The party recorded these thoughts for future reference and later found and established the Ducey series, comparable to the Shaver series but having loamy-skeletal characteristics, in certain canyons with remnants of morainal material at lower elevations.

The pack train wandered on through Woodchuck country, climbing steadily toward Crown Divide. Jack Fisher and Gordon Huntington left the train for mapping activity and photo interpretation of the locality. Hans Jenny proceeded on after the packer toward Crown Lake. Later in the day, the survey party caught up with him. He was walking, leading his horse. The party was greeted with a wry smile and informed that he "had decided to give his dogs some exercise and his horse and ass a rest!" Crown Lake camp was reached soon thereafter.

SECOND DAY—HEADWATERS OF NORTH FORK

In the early morning, camp chores were shared by all. The mules were packed, horses saddled, and plans for the day made. The trail led onward up from Crown Lake to Crown Divide 3090.9 m (10 200 ft). At the divide, the three paused to absorb the panorama of the LeConte Divide unfolding to the east. This rugged, serrated, north–south trending ridge with peaks ranging upward to 3636.4 m (12 000 ft) was the eastern boundary of the survey area. Beyond lay the high country of Kings Canyon National Park. To the north and east was the canyon of the headwaters of the North Fork of the Kings River and beyond into the rocklands of Bench Valley and the Red Mountain Basin. The latter was the objective of the previous week's traverse. To the south and east was the gentler, forested slopes of Crown Valley that drained into the Middle Fork of Kings River near Tehipite Dome. Lingering patches of snow accented protected slopes and ravines of LeConte Divide. The morning sun highlighted the glacial sculpturing of the Divide and of the uplands west of it—delicate arêtes, cirques, stretches of scoured and polished rock surfaces, and a complex of morainal debris.

Jenny elected to press on with the packer to the next camp site at Portal Lake 3121.2 m (10 300 ft) in Crown Basin. There, he wanted to explore the terrain and soils with the intent of selecting and sampling a complete tessara of soil and vegetation. Fisher and Huntington stayed behind to map this day's segment of the traverse.

Around the campfire that night, with serrated peaks of LeConte Divide outlined by starlight, conversation drifted to the glacial features seen during the day. To everyone's surprise, Jenny avowed that the features seen were the result of water erosion, and, further, that glaciers did not change the landscape but provided a blanket of protection. Little did Jack or Gordon realize at the time that Jenny was indulging in a favorite pastime of debate in which he would espouse a questionable side of an argument—and defend it admirably—just to probe the depths of understanding of a subject by his opponents. A heated, but friendly discussion ensued well into the evening. When finally concluded, considerable theory and hypotheses concerning glacial action had been exchanged and related to the terrain seen that day. Fisher and Huntington finally realized Jenny's ploy, and wryly appreciated his pedagogic method. Huntington acquired some ideas that were carried into his Ph.D. work at a later date.

THIRD DAY—THE PODOSOLS OF MCGUIRE LAKES

Sunrise over Mt. Reinstein in the LeConte Divide the next morning found the party readying to strike camp and leave for McGuire Lakes in the rockland east of Bench Valley. Before leaving, all accompanied Jenny to his tessara sample site to inspect it and to assist in completing the sampling. The soil was a shallow Regosol overlying compacted glacial flour.

The packer lead the party down into the North Fork canyon, across the tumbling waters of the river, and up the north slope in search of the markers for a "packers trail" not shown on existing maps. In time, the first marker was found and the party carefully and laboriously climbed the rocky slope of the canyon on this "short-cut" to Bench Valley. By noon, the party had crested the ridge leading up to the volcanic plug known as Blackcap 3515.2 m (11 600 ft). The ridge overlooked the morainally dammed McGuire Lakes 3030.3 m (10 000 ft). The packer pointed the way down to the next campsite on the terminal moraine between Upper and Lower McGuire Lakes. He then took the pack train on to make camp. The party spent the balance of the afternoon using the vantage point of the ridge to explore by binoculars the terrain of Bench Valley, the north-facing slopes of the North Fork of Kings River, and mapping by photo interpretation and experience of similar terrain and vegetation cover previously traversed.

Late in the day, the party made its way down to the campsite. Jenny preferred walking, leading his horse, and soon dropped behind. As Fisher and Huntington trailed beside Lower McGuire Lake, they heard a "hallo" from Jenny behind them. Huntington returned to determine the reason for the evident excitement in the hailing cry. Jenny was under a lodgepole pine on the rubbly terminal moraine backing up Lower McGuire Lake. He had carefully cut and laid aside the pine needle litter and had dug a shallow pit exposing an obvious A2 (E) horizon underlain by a weakly developed, but apparent podsolic B (Bs) horizon. Fisher and Huntington had seen occasional apparent A2 horizons in morainal and residual soils in the Hudsonian life zone elsewhere, but never in conjunction with a recognizable subsoil, nor occurring in sufficient extent to justify consideration of a new soil series. Jenny and Huntington rejoined Fisher at the packer-prepared campsite and speculated on the observation. After dinner, in the evening twilight, Jenny located another site near camp on the bouldery moraine. In the fading light, this site looked similar to the late afternoon site and heightened the excitement of the possibility of podsols in the Sierra.

FOURTH DAY—PODOSOL STUDY, NORTH FORK CANYON

The forenoon of the following day was spent in excavating several sites. The second excavation exposed a more strongly expressed podsol profile. A complete profile description, horizon sampling, tessara sampling, and an official soil-vegetation plot description of the surrounding acre was made for this site. Jenny stated that this was the first podsol he had seen in the Sierra. Fisher and Huntington affirmed that this also was true for them. It was suggested and agreed by all that the name "Blackcap" be selected for the tentative soil series name. The established Blacklock series, a podsol on California's north central coastal terraces, had already piqued Jenny's curiosity.

Subsequent field mapping by Fisher and Huntington ascertained that the Blackcap soils within the survey area were confined to granitic glacial moraine deposits above an elevation of about 2878.8 m (9500 ft) and associated with a lodgepole pine, or mixed lodgepole pine–western white pine–whitebark pine cover, but not above timberline under arctic-alpine shrub cover. In later years, Huntington found similar soils in Sequoia National Park under similar conditions.

Later that day, the party broke camp, left the McGuire Lakes area and dropped quickly by trail into the canyon of the North Fork headed toward Meadowbrook camp on the river. The eyes of the mappers, sharpened by Jenny's influence, noted the disappearance of the A2 horizons and podsolic soil features in trailside exposures of morainal veneer on the canyon slopes as they moved downslope. The more familiar light grayish AC profiles of the Maxson series soon dominated the landscape still under a cover of lodgepole pine.

The last night out on the traverse was spent in the Meadowbrook area. More philosophical discussion around the campfire enriched the party members further, including the packer who was intrigued with the flow of thoughts and ideas about the lands over which he had so often ridden.

APPENDIX 3
Brief Highlights of Hans Jenny's Life

Ronald Amundson
Department of Soil Science
University of California
Berkeley, California

Employment and Activities Record

1899	Born February 7 in Zürich, Switzerland, raised in city and country
1918–1919	Spent a year farming (French parts of Switzerland)
1923	B.S. in agriculture, Swiss Federal Technical Institute, Zürich
1927	Doctor's degree in agricultural chemistry with thesis in colloid chemistry, Swiss Federal Technical Institute
	Rockefeller Foundation Fellow, working with Professor S.A. Waksman (Nobel laureate) in New Brunswick, NJ
1928–1936	Assistant Professor of Soils, University of Missouri, Columbia, MO
1936	American citizenship
	Associate Professor of Soil Chemistry and Morphology, University of California, Berkeley, taught pedology and colloid science
1943–1949	Head, Department of Soils, University of California, Berkeley
1946	Guggenheim fellowship for study of tropical soils in Colombia, South America
1949	President, Soil Science Society of America
1954–1955	Year's leave of absence to study soils in India, Egypt, and Europe; foreign aid program and second Guggenheim fellowship
1963–1964	Fulbright lecturer, Germany
1967	Retired, but rehired for teaching for 2 yr, NSF grant for field and laboratory research
1971	United Nations consultant to Bulgaria
1977	Collaborator, Oak Ridge National Laboratory
1981	Sampled soil on Mt. Kilimanjaro, Africa
Until 1992	Wrote papers and books, conducted seminars, lead field trips, worked with graduate students, served on state committees, conducted field and laboratory research in soil science, and participated in projects for the preservation of important landscapes and for enactment of environment-enhancing legislation.
1992	Died January 9 in Oakland, CA.

Selected Honors

1930	Nitrogen research award from American Society of Agronomy
1957	Honorary Doctor of Agricultural Sciences, Liebig University, Giessen,

Copyright © 1994 Soil Science Society of America, 677 S. Segoe Rd., Madison, WI 53711, USA. *Factors of Soil Formation: A Fiftieth Anniversary Retrospective.* SSSA Special Publication 33.

Germany, "In appreciation of his fundamental work on the relations between climatic factors and soil formation, and the interactions of plants and soil colloids."

1963 Certificate of Appreciation, University of Washington, Seattle, for distinguished service as Walker Ames Professor during summer of 1963

1967 Doctor of Law, University of California, Berkeley, "Distinguished soil chemist and morphologist, whose work ushered in a new era of research into the origin and distribution of soils on the earth's surface. By analysis of functional relationships that link together the properties of soil and the environment, he introduced a quantitative approach into a previously qualitative subject. In more recent years, his method has been thoroughly established in the more general field of ecology, so that knowledge of his work is an essential part of the scientific preparation of soil scientists and ecologists. We salute today a Professor Emeritus of this University and confer on him our highest honor."

1975 Honorary Membership, the Soil Science Society of America. For contributions, dedication and service to soil science

1977 With wife Jean, Recognition Award, Soil Conservation Society of America, "For their strong support for the recognition and delineation of natural areas throughout California and their outstanding contribution toward the description, recognition and final establishment of the Pygmy Forest Ecological Staircase Natural Area in Mendocino County."

1981 With wife Jean, Fellows of the California Native Plant Society, "In recognition of continuing outstanding contributions to the botany of California native plants."

1982 With wife Jean, Certificate of Appreciation, The Nature Conservancy. "In recognition of an outstanding contribution toward the protection of the Jepson Prairie Preserve."

1983 Honor Award, Soil Conservation Society of America, "In recognition of significant contributions and achievements in land and water conservation."

1983 British Broadcasting Corporation, London, dedicates the book *Bellamy's New World* to Hans Jenny.

1984 Award of Honor, American Society of Agronomy, "For distinguished contributions to the advancement of human welfare and enhancement of California agriculture."

1984 Sparks, D.L., in "Ion activities," *Soil Science Society of America Journal* 48:514. "This paper is dedicated with immense admiration to Professor Jenny."

1984 A symposium in honor of Hans Jenny's 85th birthday was held at the University of California, Berkeley, scientists and friends from all over the country came at their own expense to present talks honoring Professor Jenny

1989 The Berkeley Citation, the highest honor which the Berkeley campus can bestow; awarded to those members of the faculty with notable and distinguished service to the University and who have gained particular distinction in their field

1992 *Orion* magazine 11(2):17, "The depth, breadth, and precision of Hans Jenny's knowledge have seldom been equalled in the life sciences of this century. Yet the quality that most set him apart was his gift for drawing together threads from realms as distant as colloid chemistry, agriculture, and microbiology to reveal a larger and previously unknown system...(We) dedicate these articles on soil to Hans Jenny: to celebrate his life, to remember him, and to honor his passion for understanding."

Publications of Hans Jenny

1. Jenny, H. 1923. A contribution to the knowledge of the calcium carbonate content of Swiss soils. (In German.) Schweiz. Landwirtsch. Z. 51:57–61.
2. Jenny, H. 1923. Foundations of animal feeding. (In German.) Schweiz. Landwirtsch. Z. 51:1176–1180.
3. Frey (-Wyseling), A., and H. Jenny. 1924 to 1925. The significance of pH in biology. (In German.) Schweiz. Z. Naturwiss. 1924–1925:23–28.
4. Jenny, H. 1925. Report on soil investigations in the Swiss National Park. (In German.) Bericht der Kommission für die Wissenschaftliche Untersuchung des Nationalparks für das Jahr 1924, Zürich.
5. Jenny, H. 1925. Interactions between fertilizer and soil. (In German.) Schweiz. Landwirtsch. Z. 53:1295–1298.
6. Jenny, H. 1925. Electrolytic sterilization of cider, system H. Jenny. *In* Süsser Most im Fass das ganze Jahr. Verlag Landschäftler A.G. Liestal, Schweiz, Germany.
7. Jenny, H. 1925. Reaktionsstudien an schweizerishen böden. Landwirtsch. Jahrb. Schweiz. 1925:261–286.
8. Jenny, H. 1926. Die alpinen Böden. *In* J. Braun-Blanquet unter Mitwirkung von H. Jenny: Vegetationsentwicklung und Bodenbildung in der alpinen Stufe der Zentralalpen. Denkschr. Schweiz. Naturforsch. Ges. 63(2):295–340.
9. Jenny, H. 1926. Anleitung zum quantitativen agrikulturchemischen Praktikum. Von G. Wiegner unter Mitwirking von H. Jenny. Velrag Gebr. Bornträger, Berlin.
10. Jenny, H. 1927. Kationen-und Anionenumtausch an Permutitgrenzflächen. Kolloidchemische Beihefte 23:428–472.
11. Wegner, G. and H. Jenny. 1927. Ueber basenaustausch an permutiten. Kolloid-Z. 42:268–272.
12. Wegner, G., and H. Jenny. 1927. On base exchange. p. 40–45. *In* Int. Congr. of Soil Sci. Abstr. of Proc. Comm. 1st, Washington, DC. (English translation.) J. Heidingsfeld, New Brunswick, NJ.
13. Jenny, H. 1928. Bemerkungen zur bodentypenkarte der Schweiz. Landwirtsch. Jahrb. Schweiz. 1928:379–383.
14. Jenny, H. 1928. Relation of climatic factors to the amount of nitrogen in soils. J. Am. Soc. Agron. 20:900–912.
15. Jenny, H. 1928. Soil investigations in the Swiss Alps. *In* Proc. Int. Congr. Soil Sci., 1st, Washington, DC. Vol. 4.
16. Jenny, H. 1929. Relation of temperature to the amount of nitrogen in soils. Soil Sci. 27:169–188.
17. Jenny, H. 1929. Klima und Klimabodentypen in Europa und in den Vereinigten Staaten von Nordamerka. Bodenk. Forsch. 1:139–189.
18. Jenny, H. 1930. Hochgebirgsböden. Blanck's Handb. Bodenlehre 3:96–118.
19. Jenny, H. 1930. The nitrogen content of the soil as related to the precipitation-evaporation ratio. Soil Sci. 29:193–206.
20. Jenny, H. 1930. An equation of state for soil nitrogen. J. Phys. Chem. 34:1053–1057.

21. Jenny, H. 1930. Gesetxmässige beziehungen swischen bodenhumus und klime. Naturwissenschaften 18:850–866.
22. Jenny, H. 1930. A study on the influence of climate upon the nitrogen and organic matter content of the soil. Missouri Agric. Exp. Stn. Res. Bull. 152:1–66.
23. Jenny, H. 1931. Soil organic matter-temperature relationship in the eastern United States. Soil Sci. 31:247–252.
24. Albrecht, W.A., and H. Jenny. 1931. Available soil calcium in relation to "damping off" of soybean seedlings. Botan Gaz. (Chicago) 42:263–278.
25. Jenny, H. 1931. Behavior of potassium and sodium during the process of soil formation. Missouri Agric. Exp. Stn. Res. Bull. 162:1–63.
26. Jenny, H. 1932. Functions between soil humus and climate. (In German.) Proc. 2nd Int. Congr. Soil Sci. Leningrad-Moscow, 1930, vol. 3, 120–131, Moscow.
27. Jenny, H. 1932. Studies on the mechanism of ionic exchange in colloidal aluminum silicates. J. Phys. Chem. 36:2217–2258.
28. Jenny, H. 1932. An ionic exchange movie. Proc. Am. Soil Surv. Assoc. Bull. 13:151–156a.
29. Jenny, H., and E.W. Cowan. 1933. The utilization of adsorbed ions by plants. Science (Washington, DC) 77:394–396.
30. Jenny, H. 1933. Soil fertility losses under Missouri conditions. Missouri Agric. Exp. Stn. Bull. 324:1–10.
31. Jenny, H., and E.W. Cowan. 1933. Über die bedeutung der im boden absorbierten kationen für das pflanzenwachstum. Z. Pflanzenernaehr. Dueng. Bodenk. A31:57–67.
32. Jenny, H. 1933. Einige praktische bodenstickstoffprobleme in den U.S.A. Schweiz. Landwirtsch. Monatsh. 9:1–15.
33. Jenny, H., and E.R. Shade. 1934. The potassium lime problem in soils. J. Am. Soc. Agron. 26:162–170.
34. Jenny, H., and C.D. Leonard. 1934. Functional relationships between soil properties and rainfall. Soil Sci. 38:363–381.
35. Hammar, C.H., and H. Jenny. 1934. Land classification as an aid in land use planning. J. Farm Econ. 16:431–433.
36. Jenny, H., and G.D. Smith. 1935. Colloid chemical aspects of clay pan formation in soil profiles. Soil Sci. 39:377–389.
37. Jenny, H. 1935. The clay content of the soil as related to climatic factors, particularly temperature. Soil Sci. 40:111–128.
38. Jenny, H., and R.F. Reitimeier. 1935. Ionic exchange in relation to the stability of colloidal systems. J. Phys. Chem. 39:593–604.
39. Jenny, H. 1935. Contribution to efficient use of Missouri lands. Missouri Agric. Exp. Stn., Columbia, MO.
40. Kelley, W.P., H. Jenny, and S.M. Brown. 1936. Hydration of minerals and soil colloids in relation to crystal structure. Soil Sci. 41:259–274.
41. Jenny, H. 1936. Simple kinetic theory of ionic exchange. I. Ions of equal valency. J. Phys. Chem. 40:501–507.
42. Kelley, W.P., and H. Jenny. 1936. The relation of crystal structure to base exchange and its bearing on base exchange in soils. Soil Sci. 41:367–382.
43. Jenny, H. 1936. Georg Wiegner. Soil Sci. 42:79–85.
44. Gieseking, J.E., and H. Jenny. 1936. Behavior of polyvalent cations in base exhange. Soil Sci. 42:273–320.

45. Jenny, H., and R. Overstreet. 1938. Contact effects between plant roots and soil colloids. Proc. Natl. Acad. Sci. 24:384-392.
46. Jenny, H. 1938. Soil colloids: Key to successful fertilization. p. 206. Vol. 815. South. Pac. Rural Press.
47. Jenny, H. 1938. Properties of colloids. Stanford Univ. Press, Stanford, CA.
48. Jenny, H., and R. Overstreet. 1939. Cation interchange between plant roots and soil colloids. Soil Sci. 47:257-272.
49. Jenny, H., and R. Overstreet. 1939. Surface migration of ions and contact exchange. J. Phys. Chem. 43:185-1196.
50. Jenny, H., and A.D. Ayers. 1939. The influence of the degree of saturation of soil colloids on the nutrient by roots. Soil Sci. 48:443-459.
51. Jenny, H., R. Overstreet, and A.D. Ayers. 1939. Contact depletion of barley roots as revealed by radioactive indicators. Soil Sci. 48:9-24.
52. Overstreet, R., and H. Jenny. 1939. Studies pertaining to the cation absorption mechanism of plants in soil. Soil Sci. Soc. Am. Proc. 4:125-130.
53. Jenny, H. 1941. Factors of soil formation. A system of quantitative pedology. 1st ed. McGraw-Hill Book Co., New York.
54. Jenny, H. 1941. Calcium in the soil: III. Pedologic relations. Soil. Sci. Soc. Am. Proc. 6:27-35.
55. Jenny, H. 1942. Base exchange in soils and in other disperse systems. Chron. Bot. 7:67-68.
56. Elgabaly, M.M., and H. Jenny. 1943. Cation and anion interchange with zinc montmorillonite clays. J. Phys. Chem 47:399-408.
57. Elgabaly, M.M., H. Jenny, and R. Overstreet. 1943. Effect of type of clay mineral on the uptake of zinc and potassium by barley roots. Soil Sci. 55:257-263.
58. Jenny, H. 1943. Alfred Smith. J. Am. Soc. Agron. 35:1060.
59. Jenny, H., A.D. Ayers, and J.S. Hosking. 1945. Comparative behavior of ammonia and ammonium salts in soils. Hilgardia 16:429-257.
60. Jenny, H. 1945. Review of E.C.J. Mohrs book: *The soils of equatorial regions*. Geogr. Rev. 35:335-336.
61. Parker, E.R., and H. Jenny. 1945. Water infiltration and related soil properties as affected by cultivation and organic fertilization. Soil Sci. 60:353-376.
62. Jenny, H. 1946. Decline of soil fertility. Spreckels Sugar Beet Bull. 10:44-45.
63. Parker, E.R., and H. Jenny. 1946. Tillage effects on soil structure. Diamond Walnut News 28:10-11.
64. Jenny, H. 1946. Adsorbed nitrate ions in relation to plant growth. J. Colloid Sci. 1:33-47.
65. Parker, E.R., and H. Jenny. 1946. Cultivation effect on irrigation. Citrus Leaves 26:6-7, 36-37.
66. Jenny, H. 1946. Arrangement of soil series and types according to functions of soil-forming factors. Soil Sci. 61:375-391.
67. Jenny, H., and Collaborators. 1946. Exploring the soils of California. p. 317-393. *In* C.B. Hutchinson (ed.) California agriculture. Univ. of California Press, Berkeley.
68. Jenny, H., and J. Vlamis. 1948. Calcium deficiency in serpentine soils as revealed by adsorbent technique. Science (Washington, DC) 107:1-3.
69. Jenny, H. 1948. Great soil groups in the equatorial regions of Colombia, South America. Soil Sci. 66:5-28.

70. Jenny, H., F.T. Bingham, and B. Padilla-Saravia. 1948. Nitrogen and organic matter contents of equatorial soils of Colombia, South America. Soil Sci. 66:173–186.
71. Jenny, H., S.P. Gessell, and F.T. Bingham. 1949. Comparative study of decomposition rates of organic matter in temperate and tropical regions. Soil Sci. 68:419–432.
72. Jenny, H. 1950. Causes of the high nitrogen and organic matter content of certain tropical forest soils. Soil Sci. 69:63–69.
73. Jenny, H. 1950. In praise of the Swiss farmer in America. The Swiss Record. (Translated by A. Senn.) Yearb. Swiss-American Hist. Soc. 2:65–66.
74. Jenny, H. 1950. Origin of soils. p. 41–61. In P.D. Trask (ed.) Applied sedimentation. John Wiley & Sons, New York.
75. Jenny, H., J. Vlamis, and W.E. Martin. 1950. Greenhouse assay of fertility of California Soils. Hilgardia 20:1–8.
76. Jenny, H., J. Vlamis, and W.E. Martin. 1950. Nutrient deficiencies in soils. Calif. Agric. 4(1):7–16.
77. Jenny, H., T.R. Nielsen, N.T. Coleman, and D.E. Williams. 1950. Concerning the measurement of pH, ion activities and membrane potentials in colloidal systems. Science (Washington, DC) 112:164–167.
78. Jenny, H. 1951. Factors of soil formation. (In Russian.) Moscow.
79. Jenny, H., O.E. Bowen, R.Z. Rollins, J.W. Vernon, W.E. ver Planck, and L.A. Wright. 1951. Minerals useful to California agriculture. Calif. Div. Mines Bull. 155:9–66.
80. Jenny, H. 1951. Contact phenomena between absorbents and their significance in plant nutrition. p. 107–132. In E. Truog (ed.) Mineral nutrition of plants. Univ. of Wisconsin Press, Madison, WI.
81. Coleman, N.T., D.E. Williams, T.R. Nielsen, and H. Jenny. 1951. On the validity of interpretations of potentiometrically measured soil pH. Soil Sci. Soc. Am. Proc. 15:106–114.
82. Jenny, H. 1951. The story of a soil scientist. Idea and experiment. Fac. Quart. 1(2):14–15.
83. Williams, D.E., and H. Jenny. 1952. The replacement of nonexchangeable potassium by various acids and salts. Soil Sci. Soc. Am. Proc. 16:216–221.
84. Jenny, H. 1953. Los grandes grupos de suelos en las regiones ecuatoriales de Colombia. Boletin Tecnico Federation Nationales de Cafeteros de Colombia 1:1–32.
85. Jenny, H, F. Bingham, and B. Padilla-Saravia. 1953. El conterindo de nitrogeno y materia organica en los suelos ecuatoriales de Colombia. Boletin Tecnico Federation Nationales de Cafeteros de Colombia 1:1–18.
86. Jenny, H, S.P. Gessel, and F.T. Bingham. 1953. Estudio comparativo sobre la velocidad de descomposicion de la materia organica en regiones tropicales y tempiados. Bol. Tecnico Fed. Nationales de Cafeteros de Columbia 1:19–39.
87. Jenny, H., F.T. Bingham, M. Llano, and J. Vlamis. 1953. Estudio sobre la fertilidad de ocho suelos Colombianos. Boletin Tecnico Federation Nationales de Cafeteros de Colombia 1:1–16.
88. Jenny, H. 1953. Book review on B.T. Shaw, *Soil Physical Conditions and Plant Growth*. J. Colloid Sci. 8:374.
89. Jenny, H. 1954. Soils. p. 67–72. In B.M. Woods (ed.) California development problems. Univ. of California.

90. Jenny, H. 1954. Phosphorus levels in California soils. p. 32–40. *In* Proc. Annu. California Fert. Conf., 2nd. Vol. 2.
91. Jenny, H. 1958. Role of the plant factor in the pedogenic functions. Ecology 39:5–16.
92. Jenny, H., E. Moreno, and J. Paul. 1958. Availability of various phosphate sources in California soils. p. 16–39. *In* Proc. Annu., Meet. Natl. Joint Comm. on Fert. Application. 1957. Natl. Plant Food Inst., Washington, DC.
93. Harridine, F., and H. Jenny. 1958. Influence of parent material and climate on texture and nitrogen and carbon contents of virgin California soils. Soil Sci. 85:235–243.
94. Jenny, H. 1958. Book review of Mohr and V. Baren's *Tropical Soils*. Ecology 39:388.
95. de Lopez-Gonzalez, J., and H. Jenny. 1958. Modes of entry of strontium into plant roots. Science (Washington, DC) 128:90–91.
96. Jenny, H. 1959. Soil as a natural resource. p. 104–109. *In* M. Huberty and W.L. Flock (ed.) Natural resources, McGraw-Hill Book Co., New York.
97. de Lopez-Gonzalez, J., and H. Jenny. 1959. Diffusion of strontium in ion-exchange membranes. J. Colloid Sci. 14:533–542.
98. Jenny, H. 1960. Podosols and pygmies: A special need for preservation. Sierra Club Bull. 45(4):8–9.
99. Jenny, H., and S.P. Raychaudhuri. 1960. Effect of climate and cultivation on nitrogen and organic matter reserves in Indian soils. Indian Council of Agric. Res., New Delhi. (p. 1–126.)
100. Glauser, R., and H. Jenny. 1960. Two-phase studies on availability of iron in calcareous soils. I. Experiments with Alfalfa plants. Agrochimica 4(4):263–278.
101. Grunes, D.L., and H. Jenny. 1960. Two-phase studies on availability of iron in calcareous soils. II. Decomposition of colloidal iron hydroxide by ion exchangers. Agrochimica 4(4):279–287.
102. Glauser, R., and H. Jenny. 1960. Two-phase studies on availability of iron in calcareous soils. III. Contact, and exchange diffusion in ionic membranes. Agrochimica 5(1):1–9.
103. Charley, J.L., and H. Jenny. 1961. Two-phase studies on availability of iron in calcareous soils. IV. Decomposition of iron oxide by roots, and Fe-diffusion in roots. Agrochimica 5(2):99–107.
104. Jenny, H. 1961. Two-phase studies on availability of iron in calcareous soils. V. Kinetics of iron transfer as conditioned by ion exchange capacity and structure of roots. Agrochimica 5(4):281–289.
105. Jenny, H. 1961. E.W. Hilgard and the birth of modern soil science. Collana della Rivista "Agrochimia." Pisa, Italy.
106. Jenny, H. 1961. Plant root-soil interactions. p. 665–694. *In* M.Y. Zarrow (ed.) Growth in living systems. Purdue Int. Symp. on Growth. June 1960. Basic Books, New York.
107. Jenny, H. 1961. Derivation of state factor equations of soils and ecosystems. Soil Sci. Soc. Am. Proc. 25:385–388.
108. Jenny, H. 1961. Comparison of soil nitrogen and carbon in tropical and temperate regions. Missouri Agric. Exp. Stn. Res. Bull. 765:1–31.
109. Jenny, H. 1961. Reflections on the soil acidity merry-go-round. Soil Sci. Soc. Am. Proc. 25:428–432.
110. Klemmedson, J.O., A.M. Schultz, H. Jenny, and H.H. Biswell. 1962. Effect of prescribed burning of forest litter on total soil nitrogen. Sol Sci. Soc. Am. Proc. 26:200–202.

111. Jenny, H., and K. Grossenbacher. 1962. Root-soil boundary zones. Calif. Agric. 16(10):7.
112. Jenny, H. 1962. Model of a rising nitrogen profile in Nile Valley alluvium, and its agronomic and pedogenic implications. Soil Sci. Soc. Am. Proc. 26:588–591.
113. Jenny, H., and K. Grossenbacher. 1963. Root-soil boundary zones as seen in the electron microscope. Soil Sci. Soc. Am. Proc. 27:273–277.
114. Jenny, H. 1963. Book review of F. Helfferich, *Ion exchange*. Soil Sci. Soc. Am. Proc. 27:iv.
115. Jenny, H. 1964. Giessens beitrag zur deutschen und internationalen bodenkunde (H. Kuron). Nachr. Ges. Hochsch. 33:29–37.
116. Jenny, H. 1965. Bodenstickstoff und seine Abhängigkeit von Zustands faktoren. Z. Pflanzenernaehr Dueng. Bodenkd. 109:97–112.
117. Jenny, H. 1965. Tessera and pedon. Soil Surv. Horiz. 6:8–9.
118. Jenny, H. 1966. Pathways of ions from soil into root according to diffusion models. Plant Soil 25:265–289.
119. Klemmedson, J.O., and H. Jenny. 1966. Nitrogen availability in California soils in relation to precipitation and parent material. Soil Sci. 102:215–222.
120. Matar, A.E., J.L. Paul, and H. Jenny. 1965. Two-phase experiments with plants growing in phosphate-treated soil. Soil Sci. Soc. Am. Proc. 31:235–237.
121. Jenny, H. 1967. Underground space frontiers. Plant Food Rev. 13:1.
122. Jenny, H. 1968. The image of soil in landscape art, old and new. Pontif. Acad. Sci. Scr. Varia 32:947–979.
123. Jenny, H, A.E. Salem, and J.R. Wallis. 1968. Interplay of soil organic matter and soil fertility with state factors and soil properties. Pontif. Acad. Sci. Scr. Varia 32:1–44.
124. Polle, E.O., and H. Jenny. 1968. Physiological activity of hydrogen and aluminum clays and resins. Soil Sci. Soc. Am. Proc. 32:528–530.
125. Jenny, H., R.J. Arkley, and A.M. Schultz. 1969. The pygmy forest-podsol ecosystem and its dune associates of the Mendocino Coast. Madroño 20:60–74.
126. Blosser, D.L., and H. Jenny. 1971. Correlations of soil pH and percent base saturation as influenced by soil forming factors. Soil Sci. Soc. Am. Proc. 35:1017–1018.
127. Polle, E.O., and H. Jenny. 1971. Boundary layer effects on ion absorption by roots and storage organs of plants. Physiol. Plant. 25:219–224.
128. Jenny, H. 1973. The pygmy forest ecological staircase. A proposal for national monument status. Berkeley, CA.
129. Jenny, H. 1976. The origin of mima mounds and hogwallows. Fremontia 4(3):27–28.
130. Jenny, H. 1980. The soil resource, origin and behavior. Springer-Verlag, New York.
131. Jenny, H. 1980. Alcohol or humus? Science (Washington, DC) 209:444.
132. Jenny, H. 1984. The making and unmaking of a fertile soil. p. 42–55. *In* W. Jackson et al. (ed.) Meeting the expectations of the land. North Point Press, San Francisco, CA.
133. Jenny, H., and K. Stuart. 1984. My friend, the soil. J. Soil Water Conserv. 39:158–161.
134. Jenny, H. 1985. History and future of soil conservation. Sierra Club Yodeler. October. p. 8–9.
135. Amundson, R., and H. Jenny. 1991. The place of humans in the state factor theory of ecosystems and their soils. Soil Sci. 151:99–109.